T0202737

Quantum Collision Theory of Nonrelativistic Particles

Reiner M. Dreizler • Tom Kirchner •
Cora S. Lüdde

Quantum Collision Theory
of Nonrelativistic Particles

An Introduction

 Springer

Reiner M. Dreizler
Institut für Theoretische Physik
Goethe Universität
Frankfurt/Main, Germany

Tom Kirchner
Department of Physics and Astronomy
York University
Toronto, ON, Canada

Cora S. Lüdde
Institut für Theoretische Physik
Goethe Universität
Frankfurt/Main, Germany

ISBN 978-3-662-65590-0   ISBN 978-3-662-65591-7   (eBook)
https://doi.org/10.1007/978-3-662-65591-7

This book is a translation of the original German 1st edition "Streutheorie in der nichtrelativistischen Quantenmechanik" by Reiner M. Dreizler, Tom Kirchner & Cora S. Lüdde, published by Springer-Verlag GmbH Germany, part of Springer Nature in 2018. The translation was done with the help of artificial intelligence (machine translation by the service DeepL.com). The present version has been revised extensively with respect to technical and linguistic aspects by the authors.

This Springer imprint is published by the registered company Springer-Verlag GmbH, DE, part of Springer Nature.
The registered company address is: Heidelberger Platz 3, 14197 Berlin, Germany

Preliminary Remarks

Scattering experiments are tools, which allow access to the quantum world. Our knowledge of molecules, atoms, nuclei and elementary particles is largely based on the analysis of such experiments. A classic example is the discovery of the atomic nucleus by elastic scattering of α-particles passing through gold foils. These experiments were carried out between 1908 and 1913 by Geiger and Marsden, following a suggestion of Rutherford. The central results were published in the English journal *Philosophical Magazine* in 1913. This publication provides a confirmation of the Rutherford scattering formula, which was derived on the basis of classical arguments and published in the 1911 issue of Phil. Mag.[1] A first example for the structural investigation of nuclei by scattering experiments is the collision of protons (denoted by p or alternatively as $^1_1$H) and lithium nuclei (e.g. the isotope $^7_3$Li). The following processes are possible in this collision:

- Elastic scattering, in which the direction of motion of the particles is changed but their kinetic energy is not

$$p + {}^7_3\text{Li} \quad \longrightarrow \quad p + {}^7_3\text{Li}.$$

- In inelastic scattering, part of the collision energy is used to excite the lithium nucleus. After the excitation, the Li nucleus returns to the ground state by emitting a photon (γ). The excitation energy determines the frequency of the photon

$$p + {}^7_3\text{Li} \quad \longrightarrow \quad p' + {}^7_3\text{Li}^* \quad \longrightarrow \quad p' + {}^7_3\text{Li} + \gamma.$$

The kinetic energy of the p-Li system is reduced after the collision.
- Additional processes in which nuclear transformations occur. They are termed reaction channels. The total energy is also conserved in these processes, but the initial energy is converted in various ways for the production of the reaction products and their kinetic energy. One can recognise the validity of charge

---

[1] The publications referred to are: H. Geiger and E. Marsden, Phil. Mag. **25**, p. 604 (1913) and E. Rutherford, Phil. Mag. **21**, p. 669 (1911).

conservation (lower index) and the conservation of nucleon number (upper index) by looking at the equations characterising the reactions

$$
\begin{aligned}
{}^{1}_{1}p + {}^{7}_{3}Li \quad &\longrightarrow \quad {}^{7}_{4}Be + {}^{1}_{0}n \\
&\longrightarrow \quad {}^{6}_{3}Li + {}^{2}_{1}H \\
&\longrightarrow \quad {}^{4}_{2}He + {}^{4}_{2}He \\
&\longrightarrow \quad {}^{6}_{2}He + {}^{1}_{1}p + {}^{1}_{1}p \\
&\longrightarrow \quad \ldots
\end{aligned}
$$

Corresponding lists can be given for the scattering of 'elementary particles'. Thus one finds, e.g., for the collision of photons and protons besides the elastic scattering

$$
\gamma + p \quad \longrightarrow \quad \gamma + p
$$

the generation of uncharged or charged π-mesons

$$
\begin{aligned}
\gamma + p \quad &\longrightarrow \quad \pi^{0} + p \\
&\longrightarrow \quad \pi^{+} + n.
\end{aligned}
$$

These examples give a rough overview of the types of reactions possible in collision processes. They are expressed in the following form: Capital letters denote the target, and the end product of the reaction. Lower case letters represent the incoming projectile and the outgoing particle (or particles).

• Elastic scattering

$$
A(a, a)A
$$

The particles are unchanged. Due to the conservation of the total energy, only their direction of motion changes.

• Inelastic scattering

$$
A(a, a^{*})A^{*}
$$

The kinetic energy of the projectile is changed, and it may also be excited but retains its identity. The target system is excited and can return to the ground state by emitting photons or disintegrate into fragments.

• Particle transfer

$$
A(a, b)B
$$

The particles before the collision differ from the particles after the collision. Thereby can

- Parts of the projectile be transferred to the target (stripping reaction)
- Parts of the target be transferred to the projectile (pickup or capture reaction)
- Particles, which are formed during the collision process, can disintegrate (breakup or fragmentation reaction)

In each case, the entrance channel consists of two particles that are initially far enough apart and move freely towards each other. When they are sufficiently close, they interact with each other. The task of the experiments is to detect and investigate the possible exit channels besides elastic scattering. This involves the identification of the exiting particles and the determination of their respective momentum and energy as a function of the initial conditions. The results obtained in this way can be used to find clues about the interaction between the particles, to explore the internal structure of the particles and, if possible, to discover particles that are not yet known. The presence of a large number of reaction channels indicates that one is confronted with a very complicated situation. The complexity depends on the collision energy and the internal structure of the collision partners, which take part in the process. For the investigation of the processes, scattering theory based on quantum mechanics is used. This book is an introduction to the scattering theory of nonrelativistic collision systems. It is intended for students, who are familiar with the basic elements of quantum mechanics.

The notion that the two collision partners move towards each other (or one of the collision partners hits a stationary particle) implies a time-dependent treatment of the collision processes. The motion of the collision partners must be represented by wave packets whose time evolution is determined by an initial value problem based on the time-dependent Schrödinger equation in the nonrelativistic case. The fact that in practice not the scattering of *one* particle by another *one* is observed, but a particle beam hits a (stationary) target (like the gold foil in the Rutherford experiment), allows a stationary formulation of the scattering theory. The task in this approach is to solve the stationary Schrödinger equation with boundary conditions that reflect the scattering situation. One can show that the time-dependent and the stationary formulations give comparable results under suitable conditions. As an approach to the scattering theory, we propose the following program:

- Foundation and extension of the stationary scattering theory for elastic scattering, the simplest type of reaction. A direct approach is the stationary treatment of the problem using the Schrödinger equation with boundary conditions characterising the incident particle beam and the outgoing scattered particles (Chap. 1).
- An alternative treatment is offered by a formulation of the theory with the aid of integral equations, in which the boundary conditions are explicitly included via the use of Green's functions (Chap. 2). Central equations are in this case the Lippmann–Schwinger equations for the scattering states and an operator equation for the T-matrix.

- A third approach, which addresses the actual time development of a single collision, is the time-dependent formulation of the problem (Chap. 3). It can be shown that a mathematically appropriate treatment of the asymptotic states in time (and in space) yields the same experimentally verifiable results (the cross sections) as the stationary formulation. The quantities that play a special role in this chapter are the Møller operators and the operator of the S-matrix, an operator closely related to the T-matrix.
- The consideration of S-matrix elements allows a unified discussion of selection rules for the results of the collision processes due to particular symmetries of a given problem, as e.g. translational or rotational symmetry (Chap. 4).
- The study of the analytic structure of the S-matrix elements as a function of complex wave numbers provides further insights into the potential scattering problem, such as a connection of the scattering solutions of the Schrödinger equation with the bound states of the corresponding energy spectrum (Chap. 5).
- It is necessary to include a discussion of partially spin-polarised beams and targets by means of a formulation using density matrices in order to analyse the results of more extensive experiments (Chap. 6).
- In Chap. 7 the multichannel situation (elastic scattering in competition with reactions such as excitation, particle transfer, etc.) is outlined. Examples of such problems are the treatment of the three-body problem on the basis of the Faddeev equations, as well as the consideration of the Distorted Wave Born Approximation (DWBA) for nucleon transfer reactions.

In all chapters, a separation of the arguments based on physics and on computational aspects is attempted. For example, in the first chapter, the basic elements of stationary potential scattering theory are covered in four sections. The fifth section contains additional information on some of the concepts as well as more detailed calculations in order to complement the preceding four sections. It is recommended to check these calculations as an exercise.

At the end of this volume one finds a list of relevant literature on collision theory. It contains

- Suggestions for further reading.
- A list of the literature cited in each of the chapters.
- Mathematical background material, that is used and referred to throughout this book, are the 'Special Functions of Mathematical Physics'. A standard work on this topic is
  M. Abramowitz and I. Stegun: Handbook of Mathematical Functions. Dover Publications, New York (1974).
  Material from this source is cited in the form 'Abramowitz/Stegun, p. yyy'. Alternative editions exist in addition to the edition of the year 1974, for instance also in the form of several open source editions.
- Additional sources dealing with 'Special Functions', are e.g.
  W. Magnus, F. Oberhettinger, R. P. Soni: Formulae and Theorems for the Special Function of Mathematical Physics. Springer Verlag, New York (1966).

- Alternatively, one can consult books specialising on one particular set of Special Functions as, e.g.,
  L. J. Slater: Confluent Hypergeometric Functions. Cambridge University Press, Cambridge (1960),
  G. N. Watson: A Treatise on the Theory of Bessel Functions. Cambridge University Press, Cambridge (first edition 1922, last edition 2018).
- Some aspects of classical mechanics are cited from the book
  R. M. Dreizler, C. S. Lüdde: Theoretical Physics 1, Theoretical Mechanics. Springer Verlag, Heidelberg (2002 and 2008)
  in the form Dreizler/Lüdde, Vol. 1, Sect. or Chap. xxx.

Contents

About the Authors

Reiner M. Dreizler Studied physics in Freiburg/Breisgau (Diploma) and at the Australian National University, Canberra (PhD). Research Associate and Assistant Professor at the University of Pennsylvania in Philadelphia. Professor of Theoretical Physics at the Goethe University, Frankfurt/Main. Until his retirement, he held the endowed S. Lyson Professorship. Member DPG, EPS, Fellow APS. Field of research: many-body systems in quantum mechanics.

Tom Kirchner Studied physics at the Goethe University, Frankfurt/M (Diploma and Doctorate). Postdoctoral researcher at York University, Toronto, Canada, and at the Max Planck Institute for Nuclear Physics, Heidelberg. After a junior professorship in Theoretical Physics at the TU Clausthal, return to York University as a full-time faculty member. Member DPG, CAP, Fellow APS. Field of work: quantum dynamics of multi-particle Coulomb systems, in particular atomic scattering processes.

Cora S. Lüdde Studied physics at the Goethe University, Frankfurt/Main (Diploma). After a family break, she returned to physics and computer science with a focus on computer arithmetic. Experience and further education in didactics. Member of DPG. Field of work: application-oriented programming, development of didactic physics software.

Elastic Scattering: Stationary Formulation—Differential Equations

1

The discussion of elastic scattering of two interacting particles can be reduced to the discussion of potential scattering of one particle with the reduced mass of the system. This is possible, if the interaction is simple enough, so that it can be written in the form of a potential field. In addition, the energy of the collision system must be low enough, so that no excitation of the particles or the generation of additional particles is possible. The discussion of elastic scattering addresses all relevant concepts of collision theory. For this reason, it is particularly suited for an introduction to the collision theory of quantum particles. The method discussed in this chapter for a nonrelativistic situation is based on the Schrödinger equation with suitable boundary conditions. This approach constitutes a direct extension of the standard introduction to quantum mechanics.

1.1 Some Basic Concepts

1.1.1 Elastic Scattering Experiments

A typical arrangement for an elastic scattering *experiment* with two quantum particles is sketched in Fig. 1.1. A thin, preferably monoenergetic particle beam is required, which is obtained by collimation of particles emerging from a source. The requirement 'monoenergetic' allows the analysis of the result of the collision process as a function of the impact energy. However, as a consequence of the uncertainty relation, the attributes "thin" versus "monoenergetic" are subject to certain limitations.

For the analysis of the incident beam, a monitor is used, in which the number of particles passing per second is registered. The beam hits a target, e.g. a foil or a gas container. Ideally, the target consists of one atomic layer. In a realistic situation, targets are used which are thin enough so that a particle in the beam is scattered by at most one target particle. If this is not the case, one has to include multiple scattering corrections.

© Springer-Verlag GmbH Germany, part of Springer Nature 2022
R. M. Dreizler et al., *Quantum Collision Theory of Nonrelativistic Particles*,
https://doi.org/10.1007/978-3-662-65591-7_1

Fig. 1.1 Experimental setup for a simple scattering experiment, schematic

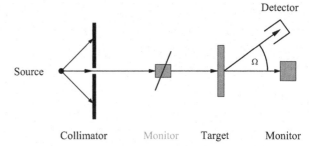

Source

Collimator   Monitor   Target   Monitor

Detector

Ω

Fig. 1.2 Detector arrangement: the detector with the aperture dΩ is located at the solid angle Ω

By means of a detector, located at distance r from the target (point) at the solid angle Ω with respect to the beam axis (Fig. 1.2), the number of scattered particles per second are counted. The detector has the smallest possible aperture dΩ. Since in practice only a very small fraction of the incident particles are deflected (in the atomic case of the order of every 10^{16}-th, in the case of scattering from nuclei every 10^{24}-th particle), it is sufficient and simpler to place the monitor in the beam behind the target (Fig. 1.1). The quantities measured are

- N $\longrightarrow$ the number of particles in the incident beam that have passed the monitor per second
- $N(\Omega)$ $\longrightarrow$ the number of particles that have arrived in the detector at the position Ω and the distance r from the target point per second (excluding the direction of the beam).

From these quantities one obtains the *differential cross section*, which is defined by

$$\left(\frac{d\sigma}{d\Omega}\right)_{exp} = \frac{r^2 N(\Omega)}{N}. \tag{1.1}$$

The factor r^2 is included in order to compensate correctly for the geometrical decrease in intensity with increasing distance of the detector from the target. The differential cross section is a function of the incident energy and the angles θ and φ which define the solid angle Ω. If the measurement is carried out for all solid angles and the measured values are summed up, one obtains the *total* cross section

$$\sigma = \int\int \left(\frac{d\sigma}{d\Omega}\right)_{exp} d\Omega = \int\int \left(\frac{d\sigma}{d\Omega}\right)_{exp} \sin\theta \, d\theta \, d\varphi. \tag{1.2}$$

1.1.2 Formulation of the Two-Particle Scattering Problem

On the *theoretical* side, the stationary treatment of the scattering problem for two interacting (structureless) particles requires the solution of the Schrödinger equation

$$\left(-\frac{\hbar^2}{2\,m_{10}}\Delta_1 - \frac{\hbar^2}{2\,m_{20}}\Delta_2 + w(r_1 - r_2)\right)\Psi(r_1, r_2) = E\,\Psi(r_1, r_2). \tag{1.3}$$

If the interaction w is conservative and Galilean invariant, the energy E is a conserved quantity. If the interaction has a finite range, so that the particles move freely at sufficiently large separations, the energy E corresponds to the sum of the (known) kinetic energies of the two particles in the initial (or the final) channel

$$E = (T_1 + T_2)_{\text{in}} = \frac{p_1^2}{2m_{10}} + \frac{p_2^2}{2m_{20}} = (T_1 + T_2)_{\text{out}} > 0.$$

Separation of relative (r) and centre of mass (R) coordinates

$$\Psi(r_1, r_2) = \psi(r)\psi_{\text{cm}}(R)$$

leads to equations for the motion of the centre of mass

$$\left(-\frac{\hbar^2}{2\,M}\Delta_R\right)\psi_{\text{cm}}(R) = E_{\text{cm}}\psi_{\text{cm}}(R)$$

and the relative motion

$$\left(-\frac{\hbar^2}{2\,\mu}\Delta_r + w(r)\right)\psi(r) = E_{\text{rel}}\psi(r).$$

The masses are the total mass $M = m_{10} + m_{20}$ and the reduced mass $\mu = m_{10}\,m_{20}/M$. For the respective energies one finds

$$E_{\text{cm}} = \frac{1}{2\,M}\left(p_1^2 + 2p_1 \cdot p_2 + p_2^2\right),$$
$$E_{\text{rel}} = \frac{1}{2\,M}\left(\frac{m_{20}}{m_{10}}p_1^2 - 2p_1 \cdot p_2 + \frac{m_{10}}{m_{20}}p_2^2\right), \tag{1.4}$$

so that the total energy is

$$E = E_{\text{cm}} + E_{\text{rel}} = T_1 + T_2.$$

The solution of the free Schrödinger equation for the motion of the centre of mass is not of interest. The differential equation for the relative motion corresponds to the differential equation for the scattering of a particle with mass m_0 by a potential $v(r)$

$$\hat{h}\psi(r) = \left(-\frac{\hbar^2}{2\,m_0}\Delta + v(r)\right)\psi(r) = E\psi(r), \tag{1.5}$$

provided one identifies[1]

$$v(r) \text{ with } w(r), \quad E \text{ with } E_{\text{rel}} \text{ and } m_0 \text{ with } \mu.$$

The Schrödinger equation (1.5) has to be solved with specified boundary conditions at the origin of the coordinate system and in the asymptotic region of space, $r \longrightarrow \infty$. The standard boundary conditions are:

- The wave function must be regular at the origin

$$\psi(r) \xrightarrow{\quad r \to 0 \quad} \text{regular.}$$

This requirement is, as for the discussion of bound states, a consequence of the structure of the kinetic energy operator, in particular the form of the centrifugal term.
- In the asymptotic region of space (asymp), one finds the incoming wave, which has passed the target and a scattered wave

$$\psi_{\text{asymp}}(r) \xrightarrow{\quad r \to \infty \quad} \psi_{\text{in}}(r) + \psi_{\text{scat}}(r).$$

For the characterisation of the incoming wave one uses in the case of structureless particles[2] a plane wave

$$\psi_{\text{in}}(r) = e^{i\,k\cdot r} \qquad \text{with } k = \frac{\sqrt{2\,m_0 E}}{\hbar}.$$

This relation characterises a beam of particles with definite energy, but by no means a *thin* beam as required for the analysis of the experiment. Rather, it describes a beam of arbitrary width which covers both the scattering centre and the detector (Fig. 1.3a). For this reason, one has to make sure that conservation of the particle

---

[1] Equation (1.5) describes either the scattering of a particle by an external potential or the relative motion of two particles in the centre of mass system. In both cases one uses, as indicated, normally the notation E, m_0 and $v(r)$.

[2] Modifications due to the spin degree of freedom or further internal degrees of freedom are discussed in Sect. 1.4.

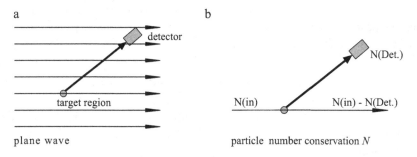

Fig. 1.3 Problems arising from the characterisation of the incoming beam by a plane wave (**a**) and the related question of particle number conservation (**b**)

number is guaranteed. The intensity of the incoming wave should be reduced after it passes the target (Fig. 1.3b). These defects can be corrected by an appropriate handling of the theoretical formulation, as will be shown below.

For a representation of the scattered part of the beam, one relies on Huygens' principle, which states that a *point-like* scattering centre is the origin of an outgoing spherical wave. As the particles are scattered by a potential (i.e. an assembly of points in a *region of space*) in the present situation, Huygens' principle has to be modified accordingly. One uses the ansatz[3]

$$\psi_{\text{scat}}(\boldsymbol{r}) = f(\theta, \varphi)\, \frac{e^{ikr}}{r}.$$

The second factor on the right-hand side of this equation represents a spherical wave emanating from the scattering centre. The first factor, the *scattering amplitude* $f(\theta, \varphi)$, represents the modification of the spherical wave. It depends on the scattering angles. This factor is obtained by a superposition of the contributions of all scattering centres. As a consequence of energy conservation in a conservative potential, one has

$$k_{\text{in}} = k_{\text{scat}} = k.$$

The momentum or wave number vectors can change their direction but not their magnitude in the case of elastic scattering.

---

[3] The plane wave, as well as the scattered part, can be normalised, e.g. with the factor $1/(2\pi)^{3/2}$. The normalisation (compare the remarks in Sect. 1.1.3) is not discussed at this point.

In most circumstances it is possible to choose a coordinate system so that the z-axis coincides with the beam direction. The asymptotic boundary condition can then be written in the form

$$\psi_{\text{asymp}}(\boldsymbol{r}) \xrightarrow{\ r \to \infty\ } e^{ikz} + f(\theta, \varphi)\, \frac{e^{ikr}}{r}. \tag{1.6}$$

The cylindrical symmetry implies in the simplest case, in which the potential depends only on the distance r, that the scattering amplitude can only depend on the polar angle

$$v(\boldsymbol{r}) \quad \to \quad v(r) \quad \Longrightarrow \quad f(\theta, \varphi) \quad \to \quad f(\theta).$$

The question, for which potentials the asymptotic form (1.6) is adequate, can not be answered in a simple fashion. For an oscillator potential with $v(r) \propto r^2$ no scattering solution is possible, for a Coulomb potential[4] with a long range ($v(r) \propto 1/r$) one can not use the ansatz (1.6). Instead of a general investigation of the correctness of the asymptotic limit suggested the following *sufficient* conditions are usually stated:

- The form (1.6) is appropriate if the range of the potential is sufficiently short. This requires a behaviour like

$$|v(\boldsymbol{r})| < \frac{c}{r^3} \quad \text{for} \quad r \longrightarrow \infty.$$

This condition is e.g. satisfied for an exponential decrease of the potential.
- If the function $v(\boldsymbol{r})$ behaves as

$$|v(\boldsymbol{r})| < \frac{c'}{r^{3/2}} \quad \text{for} \quad r \longrightarrow 0,$$

only solutions which are regular at $r \to 0$ are admitted. Potentials, which are repulsive and singular at the origin can also be handled, if the particles with an energy $E < v(r)$ are reflected at a distance r before they reach the origin (Fig. 1.4). Such potentials (hard-core potentials) do not cause any difficulties in principle (although possibly for technical reasons).
- It is required that the potentials are continuous, i.e. relatively 'smooth', over the entire region of space except for a finite number of points.

---

[4] See Sect. 1.3.

Fig. 1.4 Example of a
hard-core potential

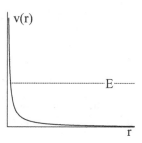

1.1.3 Connection Between Theory and Experiment

In order to connect theory and experiment, one replaces the number of particles per second by the magnitude of the *current density* j. Therefore the definition of the differential cross section from the point of view of theoretical considerations is

$$\left(\frac{d\sigma}{d\Omega}\right)_{theor} = \frac{r^2 j_{scat}}{j_{in}}. \tag{1.7}$$

If one calculates the current density j_{scat} with the scattered part of the asymptotic wave function, the excessive width of the incident beam plays no role. Particles, which are not scattered do not contribute to the differential cross section. The cross section as given in (1.7) is independent of the overall normalisation of the asymptotic wave function, i.e. independent of the strength of the incident beam. It also agrees with (1.1). For the differential cross section one finds (details in Sect. 1.5.1) with (1.6)

$$\left(\frac{d\sigma}{d\Omega}\right)_{theor} = |f(\theta, \varphi)|^2. \tag{1.8}$$

The total cross section is

$$\sigma = \iint |f(\theta, \varphi)|^2 \, d\Omega. \tag{1.9}$$

The scattering amplitude f turns out to be a central quantity of potential scattering theory.

The objection that the asymptotic form does not reflect the attenuation of the incident beam is overcome in the case of elastic scattering by invoking particle number conservation. Any particle which enters a sufficiently large sphere around the scattering centre must also leave this sphere. This requirement leads to the statement

$$\iint_{large\,sphere} \boldsymbol{j} \cdot d\boldsymbol{f} = 0.$$

If one evaluates this condition with the asymptotic wave function (1.6), one finds (details in Sect. 1.5.2) in leading order in $1/r$

$$\sigma = \frac{4\pi}{k}\,\mathrm{Im}\,f(0, \varphi) \quad \longrightarrow \quad \frac{4\pi}{k}\,\mathrm{Im}\,f(0). \tag{1.10}$$

The scattering amplitude in the forward direction ($\theta = 0$) does not depend on the polar angle φ. This relation (1.10) between the total cross section and the imaginary part of the scattering amplitude is known as the *optical theorem*. The theorem must be satisfied by any physically meaningful scattering amplitude. It guarantees that the contribution of the beam and the contribution of the scattered wave in the asymptotic wave function in forward direction (behind the target) interfere in such a way, that particle number conservation, and hence the expected intensity loss of the beam, is described correctly. From a theoretical point of view, the theorem states that the scattering amplitude is necessarily a complex quantity whose real and imaginary parts are linked by (1.10). From a practical point of view, it offers the possibility to calculate the total cross section with (1.10) rather than with (1.9).

In the next sections, the question has to be answered how the calculation of the scattering amplitude for a specified potential can be accomplished. If one relies on the differential equation (1.5)

$$\hat{h}\psi(\boldsymbol{r}) = \left(-\frac{\hbar^2}{2m_0}\Delta + v(\boldsymbol{r})\right)\psi(\boldsymbol{r}) = E\psi(\boldsymbol{r}),$$

the standard method for sufficiently short-ranged potentials is the partial wave expansion. As an alternative, on can replace the differential equation (1.5) including the boundary condition (1.6) by an integral equation. It turns out that the treatment of the problem by means of an integral equation is more suitable both for applying approximations as well as for dealing with multichannel situations. In the case of a Coulomb potential, an exact solution of the differential equation (1.5) with explicit consideration of the asymptotic limit is required, as the boundary condition (1.6) can not be satisfied.

1.1.4   The Differential Cross Section in the Laboratory System and in the Centre of Mass System

A technical point needs to be clarified before the theory can be developed further. The differential cross section can be calculated either with respect to the laboratory system (characterised by lower case letters) or with respect to the centre of mass system (characterised by upper case letters) of the two colliding particles. The experiment takes place in the laboratory. In the theoretical treatment the centre of mass system is usually preferred. It is therefore necessary to relate the statements about scattering angles and cross sections in the two systems. This requires a

Fig. 1.5 Scattering angles in
the laboratory system (black)
and in the centre of mass
system (grey)

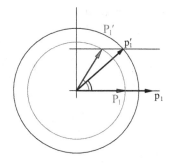

transformation for the angles between the momentum vectors, e.g. of particle 1, before the collision p_1 and after the collision p_1' in the laboratory system and the corresponding angles between the momentum vectors P_1 and P_1' in the centre of mass system. In addition, a transformation between the corresponding differential cross sections is needed. These transformations have a relatively simple form if one can assume that one of the particles is bound in such a way in the target, that it can be considered to be at rest (from a practical point of view) before the impact. The situation is illustrated in Fig. 1.5. If the condition $p_2 = 0$ applies, the momentum vectors before the collision (P_1 and p_1) are parallel and, as a result of energy conservation, the magnitude of the vector P_1' is equal to the magnitude of P_1

$$|P_1'| = |P_1|.$$

In addition, the difference of the vectors p_1' and P_1' is parallel to p_1

$$(p_1' - P_1') \ \parallel \ p_1.$$

With the help of an explicit consideration of the collision kinematics (details are presented in Sect. 1.5.3) one finds the relation

$$\cos\theta_{\text{lab}} = \frac{\dfrac{m_{10}}{m_{20}} + \cos\theta_{\text{cm}}}{\left[\left(\dfrac{m_{10}}{m_{20}}\right)^2 + 2\,\dfrac{m_{10}}{m_{20}}\cos\theta_{\text{cm}} + 1\right]^{1/2}}$$

between the scattering angles in the laboratory system θ_{lab} and in the centre of mass system θ_{cm}. The differential cross section in the laboratory system

$$\left(\frac{d\sigma}{d\Omega}\right)_{\text{lab}} = \left(\frac{d\sigma}{d\Omega_{\text{lab}}}\right)$$

is therefore related to the corresponding cross section in the centre of mass system by

$$
\begin{aligned}
\left(\frac{d\sigma}{d\Omega}\right)_{\text{lab}} &= \left(\frac{d\sigma}{d\Omega}\right)_{\text{cm}} \left(\frac{d\Omega_{\text{cm}}}{d\Omega_{\text{lab}}}\right) = \left(\frac{d\sigma}{d\Omega}\right)_{\text{cm}} \left| \left(\frac{d\cos\theta_{\text{cm}}}{d\cos\theta_{\text{lab}}}\right) \right| \\
&= \left(\frac{d\sigma}{d\Omega}\right)_{\text{cm}} \frac{\left[1 + 2(m_{10}/m_{20})\cos\theta_{\text{cm}} + (m_{10}/m_{20})^2\right]^{3/2}}{[1 + (m_{10}/m_{20})\cos\theta_{\text{cm}}]}.
\end{aligned}
\tag{1.11}
$$

It follows that

$$
\sigma = \int\int \left(\frac{d\sigma}{d\Omega}\right)_{\text{lab}} d\Omega_{\text{lab}} = \int\int \left(\frac{d\sigma}{d\Omega}\right)_{\text{cm}} d\Omega_{\text{cm}}.
$$

The total cross section is, of course, independent of the choice of the coordinate system.

1.2   The Partial Wave Expansion

A direct method for solving the boundary value problem characterised by (1.5) and (1.6) is the *partial wave expansion*. One expands the scattering wave function in terms of spherical harmonics and radial functions

$$
\psi(k, r) = \sum_{l=0}^{\infty} \sum_{m=-l}^{+l} C_{l,m}(k) F_l(k, r) Y_{lm}(\Omega).
\tag{1.12}
$$

The quantum numbers for the angular part represent the magnitude of the angular momentum ($l = 0, 1, 2, \ldots, \infty$) and the projection of the angular momentum vector onto the z-axis, the azimuthal quantum number m

$$
m = -l, -l+1, \ldots, 0, l-1, l.
$$

The quantity k is the magnitude of the vector for the wave number, which is determined by the energy E

$$
k = \frac{\sqrt{2m_0 E}}{\hbar}.
$$

The operator for the kinetic energy separates in angular and radial coordinates

$$
\begin{aligned}
\hat{T} &= \hat{T}_r + \hat{T}_\Omega \\
&= -\frac{\hbar^2}{2m_0}\left(\frac{\partial^2}{\partial r^2} + \frac{2}{r}\frac{\partial}{\partial r}\right) + \frac{\hat{l}^2}{2m_0 r^2}.
\end{aligned}
$$

It contains the operator for the square of the angular momentum

$$\hat{l}^2 = -\hbar^2 \frac{1}{\sin\theta} \frac{\partial}{\partial\theta} \left(\sin\theta \frac{\partial}{\partial\theta} \right) - \frac{\hbar^2}{\sin^2\theta} \frac{\partial^2}{\partial\varphi^2}.$$

The solution of the eigenvalue problem for the total angular part with the condition of a square-integrable (continuous, finite and unique) solution

$$\hat{l}^2 Y(\Omega) = \lambda Y(\Omega)$$

yields the eigenvalues quoted above $\lambda \to \lambda_l = \hbar^2 l(l+1)$ and as eigenfunctions the orthonormal spherical harmonics

$$\int \int d\Omega \, Y_{lm}^*(\Omega) Y_{l'm'}(\Omega) = \delta_{ll'} \delta_{mm'}.$$

A separation of the angular part

$$Y(\Omega) = P(\cos\theta) \, S(\varphi)$$

leads to the exponential functions

$$S(\varphi) \to S_m(\varphi) = e^{im\varphi}/\sqrt{2\pi}$$

as solutions of the eigenvalue problem

$$\hat{l}_z S_m(\varphi) = -i\hbar \frac{\partial}{\partial\varphi} S_m(\varphi) = m\hbar S_m(\varphi)$$

and the associated Legendre functions $P_l^m(\cos\theta) \equiv P_l^m(y)$. These functions are the regular solutions of the differential equation

$$(1 - y^2)\frac{d^2 P_l^m(y)}{dy^2} - 2y\frac{d P_l^m(y)}{dy} + \left(l(l+1) - \frac{m^2}{(1 - y^2)} \right) P_l^m(y) = 0.$$

A special case of the associated Legendre functions are the Legendre polynomials $P_l(y)$ with

$$P_l(y) = P_l^0(y).$$

In the next section, the case of a central potential $v(r) = v(r)$ will be discussed. The scattering of a particle by potentials without central symmetry is addressed in Sect. 3.5.2.

1.2.1   Partial Wave Expansion in the Case of Cylindrical Symmetry

In the case of the scattering of a particle by a central potential, the initial direction of the beam can be identified with the z-direction. As a result of the symmetry, there is no dependence on the azimuthal angle φ, so that an expansion in terms of Legendre polynomials is sufficient. This indicates, that the partial wave expansion of the scattering solution of the Schrödinger equation in the case of cylindrical symmetry constitutes an analysis of this function with respect to the magnitude of the angular momentum.

One finds in the literature variants for the expansion of the scattering wave function

$$\psi(k, r) = \sum_{l=0}^{\infty} F_l(k, r) P_l(\cos\theta) \tag{1.13}$$

and

$$\psi(k, r) = \sum_{l=0}^{\infty} \frac{R_l(k, r)}{kr} P_l(\cos\theta). \tag{1.14}$$

The angle θ is the angle between the direction of the initial beam

$$\boldsymbol{p}_{\text{in}} = \hbar \boldsymbol{k}_{\text{in}} = \hbar k \boldsymbol{e}_z$$

and the (variable) direction under which the detector is placed

$$\boldsymbol{p}_{out} = \hbar k \boldsymbol{e}_\theta.$$

Substituting (1.13) into the Schrödinger equation (1.5) gives, due to the orthogonality of the Legendre polynomials, a set of differential equations for the radial wave functions $F_l(k, r)$

$$\frac{1}{r^2} \frac{\text{d}}{\text{d}r} \left(r^2 \frac{\text{d}F_l(k, r)}{\text{d}r} \right) + \left[k^2 - \frac{l(l+1)}{r^2} - U(r) \right] F_l(k, r) = 0 \tag{1.15}$$

or in detail

$$\frac{\text{d}^2 F_l(k, r)}{\text{d}r^2} + \frac{2}{r} \frac{\text{d}F_l(k, r)}{\text{d}r} + \left[k^2 - \frac{l(l+1)}{r^2} - U(r) \right] F_l(k, r) = 0,$$

$$U(r) = \frac{2m_0}{\hbar^2} v(r), \qquad l = 0, 1, 2, \ldots.$$

For the ansatz (1.14) one obtains the somewhat simpler equation

$$\frac{d^2 R_l(k,r)}{dr^2} + \left[k^2 - \frac{l(l+1)}{r^2} - U(r) \right] R_l(k,r) = 0. \tag{1.16}$$

This equation will be discussed in the following. The solutions of the two differential equations are connected in a simple way

$$R_l(k,r) = kr F_l(k,r).$$

It is convenient to divide the space into three concentric spherical regions around the scattering centre at $r = 0$, if the potential energy, or the potential $U(r)$, falls off fast enough. One then defines

- (G1): the scattering region S, in which the potential is not equal to 0, $U(r) \neq 0$, with the wavefunction ψ_{scat},
- (G2): an intermediate region I, where $U(r) = 0$, but (at least from a practical point of view) the asymptotic condition, which is only valid for $r \to \infty$ does not yet apply, with ψ_I,
- (G3): the asymptotic region, where $U(r) = 0$ and the boundary condition (1.6) applies, with ψ_{asymp}.

The radial wave functions (1.13) or (1.14) in the region S are exact solutions of the Schrödinger equations (1.15) or (1.16) for the specified potential. The general solution of the free Schrödinger equation is valid in the intermediate region I.

The radial wave functions in Eqs. (1.13) and (1.14) and their first derivatives have to be connected in the two inner regions at a suitable radius. Instead of doing this for the wave function and the first derivative separately, it is often sufficient to consider the logarithmic derivative

$$\frac{1}{R_l^{G1}(k,r)} \frac{dR_l^{G1}(k,r)}{dr} \bigg|_{r_{(G_1,G_2)}} = \frac{1}{R_l^{G2}(k,r)} \frac{dR_l^{G2}(k,r)}{dr} \bigg|_{r_{(G_1,G_2)}}, \tag{1.17}$$

as the normalisation of the scattering wave functions does not play a role for this connection. If the solution ψ_I is then connected to the asymptotic solution ψ_{asymp}, the solution in the scattering region is matched to the asymptotic boundary condition.

The first step is the determination of the general solution of the free Schrödinger equation in the intermediate region. It can be expressed in the form of a partial wave expansion with the general solutions of the radial equations, the differential equations (1.16) with $U(r) = 0$. The substitution $x = kr$ leads to the differential equation for the Bessel-Riccati functions (details in Sect. 1.5.4)

$$\frac{d^2 R_l(x)}{dx^2} + \left(1 - \frac{l(l+1)}{x^2} \right) R_l(x) = 0.$$

The regular solutions, which behave as x^{l+1} at the origin, are denoted by $u_l(x)$. The irregular solutions,[5] which behave near the origin as $1/x^l$, are the functions $v_l(x)$. The general solution of the Schrödinger equation in the intermediate region I is therefore e.g.

$$\psi_I(r) = \sum_{l=0}^{\infty} \frac{[A_l u_l(x) + B_l v_l(x)]}{x} P_l(\cos\theta), \qquad (x = kr). \qquad (1.18)$$

In the asymptotic region $x \longrightarrow \infty$ one finds for the solutions[6] of the radial equations (1.16)

$$u_l(x) \longrightarrow \sin(x - l\pi/2), \qquad v_l(x) \longrightarrow \cos(x - l\pi/2).$$

The next step is the implementation of the boundary condition (1.6) in the form (1.14). One uses the expansion of the plane wave in terms of Bessel functions as

$$e^{ikz} = \sum_{l=0}^{\infty} i^l (2l + 1) \frac{u_l(x)}{x} P_l(\cos\theta),$$

or with the asymptotic form of the Riccati functions $u_l(kr)$

$$e^{ikz} \xrightarrow{r \to \infty} \frac{1}{kr} \sum_{l=0}^{\infty} i^l (2l + 1) \sin(kr - l\pi/2) P_l(\cos\theta),$$

as well as the expansion of the scattering amplitude $f(\theta)$ in terms of *partial scattering amplitudes* f_l

$$f(\theta) = \sum_{l=0}^{\infty} f_l \, P_l(\cos\theta). \qquad (1.19)$$

---

[5] The corresponding solutions of the differential equation (1.15) are the spherical Bessel and Neumann functions

$$j_l(x) = \frac{u_l(x)}{x} \quad \text{and} \quad n_l(x) = \frac{v_l(x)}{x}.$$

[6] Watch out: Alternative signs for the regular or the irregular solutions of the homogeneous differential equations (1.16) are often used.

The partial amplitudes f_l are constant quantities. They are ultimately determined by solution of the Schrödinger equation in the scattering region. The form of (1.14), adjusted to the boundary condition, is therefore

$$\psi_{asymp}(\boldsymbol{r}) \xrightarrow{r \to \infty} \sum_{l=0}^{\infty} \left\{ \frac{i^l}{k}(2l+1)\sin(kr - l\pi/2) + f_l\, e^{i k r} \right\} \times \frac{P_l(\cos\theta)}{r}.$$

(1.20)

The function (1.20) is the partial wave expansion of the asymptotic wave function (1.6). It should be noted, that this function is not a solution of the free Schrödinger equation.

1.2.2 The Phase Shift

If one uses a normalisation factor N_l and a *phase shift* δ_l instead of the integration constants A_l and B_l in (1.18)

$$A_l = N_l \cos\delta_l, \qquad B_l = N_l \sin\delta_l$$

as well as the inverse

$$N_l = [A_l^2 + B_l^2]^{1/2}, \qquad \tan\delta_l = \frac{B_l}{A_l},$$

one obtains for the wave function (1.18) in the intermediate region, after rewriting the trigonometric functions with the addition theorem,

$$\psi_I(\boldsymbol{r}) \xrightarrow{r \to \infty} \frac{1}{r} \sum_{l=0}^{\infty} \frac{N_l}{k} \sin(kr - l\pi/2 + \delta_l) P_l(\cos\theta).$$

(1.21)

Application of the condition (1.17) for the connection with the radial components of the functions (1.20) and (1.21) gives a representation of the partial scattering amplitudes in terms of the phase shifts (the detailed calculation is found in Sect. 1.5.5)

$$f_l = \frac{(2l+1)}{k}\sin\delta_l\, e^{i\delta_l} = \frac{(2l+1)}{2ik}\left(e^{2i\delta_l} - 1\right).$$

(1.22)

The connection of the wave functions themselves provides the normalisation factor

$$N_l = (2l+1)i^l e^{i\delta_l}.$$

(1.23)

A comparison of (1.20) and (1.21) shows that the phase shifts represent a shift of the phases of the actual solution in the asymptotic region with respect to the plane wave part of the asymptotic wave function. From the relation (1.22) one gleans the statement that a partial wave with the quantum number l does not contribute to the scattering if the corresponding phase shift is zero.

The total cross section can be represented by the partial scattering amplitudes or by the phase shifts. It is

$$
\sigma = \sum_{l,l'=0}^{\infty} f_l^* f_{l'} \int_{-1}^{1} d(\cos\theta)\, P_l(\cos\theta) P_{l'}(\cos\theta) \int_{0}^{2\pi} d\varphi
$$

$$
= 4\pi \sum_{l=0}^{\infty} \frac{|f_l|^2}{(2l+1)} = \frac{4\pi}{k^2} \sum_{l=0}^{\infty} (2l+1)\sin^2\delta_l.
$$

(1.24)

The optical theorem (1.10) is satisfied, as the relation

$$
\frac{4\pi}{k}\mathrm{Im}(f(0)) = \frac{4\pi}{k}\sum_{l=0}^{\infty}\mathrm{Im}(f_l)P_l(1) = \frac{4\pi}{k^2}\sum_{l=0}^{\infty}\left(\frac{(2l+1)}{2}(1-\cos 2\delta_l)\right)
$$

$$
= \frac{4\pi}{k^2}\sum_{l=0}^{\infty}(2l+1)\sin^2\delta_l
$$

holds.

The actual task, which has to be faced after these preparations, is the solution of the radial equation (e.g. (1.16)) in the scattering region

$$
\frac{d^2 R_l(r)}{dr^2} + \left[k^2 - \frac{l(l+1)}{r^2} - U(r)\right] R_l(r) = 0,
$$

$$
U(r) = \frac{2m_0}{\hbar^2} v(r),
$$

(1.25)

with the boundary condition at the origin $R_l(r) \xrightarrow{r\to 0} r^{l+1}$ as well as a connection of the partial solutions to the solutions in the intermediate region.

The region of space, in which the actual solution must be obtained, is reduced by the use of the partial wave expansion. In addition, one expects that for a short-range potential (range $R = R_U$) only partial waves with $l < kR$ contribute. In this case only a finite number of ordinary differential equations have to be considered. On the other hand, only a few examples are known, for which an analytical determination of the solution in the scattering region is possible. In general, one has to rely on the use of numerical methods.

1.2.3   An Example: Scattering by a Spherical Potential Well

One of the examples for which an analytical solution can be given, is the spherical potential well or the spherical potential barrier

$$v(r) = \begin{cases} \pm v_0 & \text{for} \quad r \le R, \\ 0 & \text{for} \quad r > R, \end{cases} \quad \text{with} \quad v_0 > 0.$$

If one assumes that $v_0 > 0$, then $-v_0$ describes a potential well. This simple example can be used to gain some insight into the structure of the quantum mechanical scattering problem. If one defines the effective wave number K in the scattering region by

$$k^2 - U(r) = k^2 \mp \frac{2m_0}{\hbar^2}v_0 = K^2,$$

where the upper sign applies to the case of a barrier, one recognises in (1.25) the differential equation for the Bessel-Riccati functions

$$R_l''(r) + \left(K^2 - \frac{l(l+1)}{r^2} \right) R_l(r) = 0.$$

This means, this $R_l(r)$ is either $u_l(Kr)$ or $v_l(Kr)$, or a linear combination of the two functions. As a consequence of the boundary condition at the origin, only the regular solutions, the Bessel-Riccati functions $u_l(Kr)$ are admissible. Evaluation of the logarithmic matching condition at the point $r = R_{SI} = R$ for the transition from the scattering region (wave number K) to the intermediate region (wave number k)

$$\frac{u_l'(Kr)}{u_l(Kr)}\bigg|_R = \frac{A_l u_l'(kr) + B_l v_l'(kr)}{A_l u_l(kr) + B_l v_l(kr)}\bigg|_R$$

results in

$$\tan \delta_l = \frac{B_l}{A_l} = \frac{u_l'(KR)u_l(kR) - u_l(KR)u_l'(kR)}{u_l(KR)v_l'(kR) - u_l'(KR)v_l(kR)}. \tag{1.26}$$

The scattering problem is solved. For a given initial energy (characterised by the wave number k) and the given potential (characterised by the parameters $\pm v_0$ and R), one can calculate the phase shifts with (1.26) and with (1.22) the partial scattering amplitudes. Thus one finds, e.g. for the partial wave with $l = 0$ the functions

$$u_0(kr) = \sin kr, \qquad v_0(kr) = \cos kr$$

and the result

$$\tan \delta_0 = \frac{k \sin KR \cos kR - K \cos KR \sin kR}{k \sin KR \sin kR + K \cos KR \cos kR}. \tag{1.27}$$

From the result (1.26) and (1.27) one can extract some properties of the scattering solutions, which are also valid (if applied with some caution) for other short-range potentials:

- For low collision energies $k^2 < |2m_0v_0/\hbar^2|$, the wave function and its derivative in the scattering region are, to a good approximation, independent of the wave number k. If one writes in abbreviation

$$\frac{u_l'(KR)}{u_l(KR)} \xrightarrow{k\to 0} \beta_l$$

and uses the limiting values[7]

$$u_l(kR) \xrightarrow{kR\to 0} \frac{(kR)^{(l+1)}}{(2l+1)!!} \quad \text{and} \quad v_l(kR) \xrightarrow{kR\to 0} \frac{(2l-1)!!}{(kR)^l},$$

one finds

$$\tan \delta_l(k) \approx \delta_l(k) = \left[\frac{l+1-\beta_l R}{l+\beta_l R}\right]\frac{(kR)^{(2l+1)}}{(2l-1)!!(2l+1)!!}.$$

The phase shifts behave for low collision energies (compared to the potential energy in the inner region) as

$$\delta_l(k) \propto k^{2l+1} \propto E^{l+1/2}.$$

This result shows, that, for low energies (more precisely for $kR < 1$), the contributions of the partial waves with low angular momentum values dominate. The reason is the presence of the centrifugal barrier.

- Considering the differential cross section

$$\frac{d\sigma}{d\Omega} = \sum_{l,l'=0}^{\infty} f_l^* f_{l'} P_l(\cos\theta) P_{l'}(\cos\theta)$$

---

[7] $(2l+1)!! = 1 \cdot 3 \cdot \ldots \cdot (2l+1)$.

for a given energy as a function of the scattering angle, one can observe: At sufficiently low energies, only S-waves ($l = 0$) contribute and one finds

$$\delta_0 \neq 0, \quad \delta_l \approx 0 \quad \text{for } l \geq 1 \quad \longrightarrow \quad \frac{d\sigma}{d\Omega} \approx |f_0|^2.$$

The differential cross section is independent of the scattering angle for such energy values. The scattering is isotropic. If the energy is increased, P-waves ($l = 1$) come into play. The differential cross section

$$\frac{d\sigma}{d\Omega} \approx |f_0|^2 + (f_0^* f_1 + f_0 f_1^*) \cos\theta + |f_1|^2 \cos^2\theta$$

now shows a characteristic interference pattern. This pattern can be represented either in the form of a polar plot (Fig. 1.6a) or more directly as a function of θ (Fig. 1.6b). In the polar representation, the contribution of S-waves is described by a circle around the coordinate origin. If P-waves are included, the scattering in the beam direction ($\theta = 0$) is somewhat increased. In the picture on the right-hand side the same behaviour is illustrated directly. As a result of the symmetry with respect to the position $\theta = \pi$, the function $f(\theta)$ is only shown for the interval $0 \leq \theta \leq \pi$. The constant S-wave contribution is modified by the contribution of the P-wave. The interference structure allows some conclusions concerning the form of the potential energy in the scattering region.

For S-waves, as well as for other partial waves with low angular momentum, one can follow the variation of the phase shifts over the entire range of energies or wave numbers in detail.

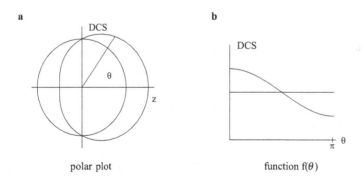

Fig. 1.6 Potential well: differential cross section (DCS), sum of the contributions with $l = 0$ and $l = 1$ (S- and P-waves). (**a**) as polar plot. (**b**) as a function of θ

Detailed Examination of S-Wave Scattering

For high collision energies one has $k^2 > |2m_0v_0/\hbar^2|$ and therefore $K \approx k$: In this case, one can use to a good approximation the addition theorem for the trigonometric functions and obtain

$$\tan \delta_0 \approx \frac{\sin(K-k)R}{\cos(K-k)R} = \tan(K-k)R.$$

Due to the ambiguity of the arc tangent function the phase shift itself is given by

$$\delta_0 \approx (K-k)R + n\pi.$$

For high energies one can use for each partial wave, as $k, K \longrightarrow \infty$, the asymptotic form of the Bessel-Riccati functions

$$u_l(x) \xrightarrow{x \to \infty} \sin\left(x - \frac{l\pi}{2}\right), \qquad v_l(x) \xrightarrow{x \to \infty} \cos\left(x - \frac{l\pi}{2}\right)$$

and combine the numerator and denominator in (1.26) for $K \approx k$ with the addition theorem, so that one obtains in this limiting case for each l

$$\tan \delta_l \longrightarrow \tan[(K-k)R].$$

One usually specifies the phase shifts at high energies by the choice $n = 0$ in order to deal with the ambiguity of the arc tangent function. If one expands the root in

$$\delta_0 \approx kR \left\{ \left[1 \mp \frac{2m_0v_0}{\hbar^2 k^2} \right]^{1/2} - 1 \right\},$$

one obtains

$$\delta_0 \xrightarrow{kR \to \infty} \mp \frac{m_0}{\hbar^2} \frac{v_0 R}{k}.$$

The upper sign applies in the case of the barrier. For both the well and the barrier, the phase shift of the S-wave decreases according to $1/k$ (resp. $1/\sqrt{E}$). Comparison of the functions $\sin kr$ and $\sin(kr + \delta_0)$ shows that the wave function for a potential well is drawn into the well. The wave function for the barrier is pushed out of the potential region. These statements show in simple form how information about the interaction of two particles can be obtained from the phase shifts.

In the case of the **barrier** one finds the following ranges for the wave numbers k and K

$$k^2 > (2m_0 v_0)/\hbar^2 = U_0 \quad \rightarrow K = [k^2 - U_0]^{1/2} > 0 \text{ real,}$$
$$k^2 = U_0 \qquad\qquad\qquad \rightarrow K = 0,$$
$$k^2 < U_0 \qquad\qquad\qquad \rightarrow K = [k^2 - U_0]^{1/2} \qquad \text{imaginary.}$$

The calculation of the phase shifts therefore requires the consideration of the three possible cases:

- In the range of high wave numbers, one can directly apply the formula (1.27).
- For the value $K = 0$ one finds by expansion

$$\tan \delta_0 (k^2 = U_0) = \frac{kR \cos kR - \sin kR}{kR \sin kR + \cos kR}.$$

- For low wave numbers one can use the substitution of the trigonometric functions by hyperbolic functions

$$\cos i x = \cosh x, \quad \sin i x = i \sinh x, \ (x \text{ real})$$

and obtain

$$\tan \delta_0 = \frac{k \sinh(|K|R) \cos kR - |K| \cosh(|K|R) \sin kR}{k \sinh(|K|R) \sin kR + |K| \cosh(|K|R) \cos kR}.$$

The energy dependence of the function $\delta_0(E)$ (instead of the dependence on the wave number) in the range

$$0 \leq E \leq 30 \left[\frac{2m_0}{\hbar^2} \right]$$

is illustrated in Fig. 1.7. The behaviour for $E \rightarrow 0$ and $E \rightarrow \infty$ corresponds, as shown explicitly in Figs. 1.8a, b, to the statements in the previous, more general discussion. The function $\delta_0(E)$ is for all energy- or k-values negative and finite. It is limited to the range of values $0 \leq \delta_0 \leq -\pi/2$.

For a **potential well** one has $K = [k^2 + U_0]^{1/2}$. The effective wave number is therefore positive definite and real. The formula (1.27) or the equivalent

$$\tan \delta_0 = \frac{k \tan KR - K \tan kR}{K + k \tan KR \tan kR} \tag{1.28}$$

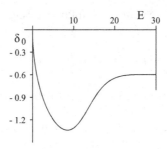

Fig. 1.7 The S-wave phase shift for a potential barrier in the energy range $0 \leq E \leq 30$ in units of $(2m_0/\hbar^2)$

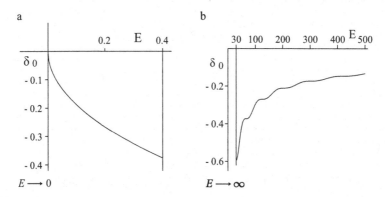

Fig. 1.8 S-wave phase shift for a potential barrier at small (**a**) and large (**b**) collision energies (energy scale as in Fig. 1.7)

can be tabulated directly. One finds the following results in this case:

1. If the potential parameters satisfy the relation

$$\bullet \quad [2m_0v_0]^{1/2}\frac{R}{\hbar} < \frac{\pi}{2},$$

there exists[8] no bound state in the well. For these parameter values one finds a behaviour for the phase shift $\delta_0(E)$, which is illustrated in Fig. 1.9a (low energy) and in 1.9b (larger energy range). The phase shift is positive in the whole energy range, first growing from 0 to a maximum value and then slowly decreasing back to the value 0. Figure 1.10 shows the analogous behaviour of the function $\tan \delta_0$ in the same energy range as in Fig. 1.9b.

---

[8] The discussion of bound states of a particle in a spherical square well potential is a standard topic in introductory texts on quantum mechanics. A graphical method for the determination of the

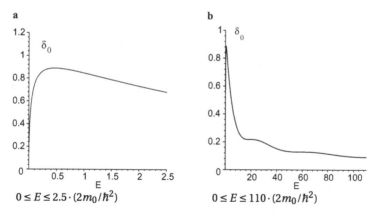

Fig. 1.9 The dependence of the S-wave phase shift for a potential well with $[2m_0v_0]^{1/2}R/\hbar <$ $\pi/2$ for small (**a**) and large (**b**) values of the collision energy

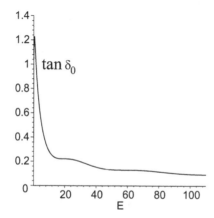

Fig. 1.10 The function $\tan\delta_0(E)$ corresponding to Fig. 1.9 b

2. For a potential characterised by

$$\bullet \quad \frac{\pi}{2} \leq [2m_0v_0]^{1/2}\frac{R}{\hbar} < \frac{3\pi}{2}$$

one bound state exists in the potential well and one obtains the behaviour shown in Fig. 1.11a for the tangent of the phase shift as a function of the collision energy. The function $\tan\delta_0$ is negative for energies below a value E_S, which is determined by the potential parameters. After a step from $-\infty$ to $+\infty$ the

bound (and virtual) states of the square well problem is described in Sect. 5.4. Consult one of the introductory texts of quantum mechanics or the Internet to find a more extensive treatment.

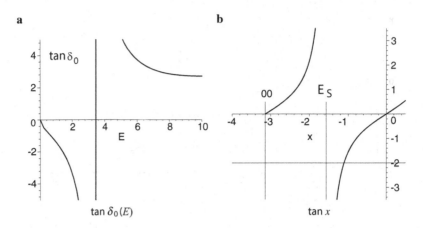

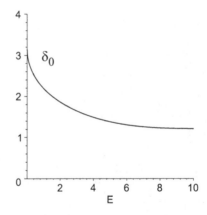

Fig. 1.11 S-wave scattering by a potential well with one bound state (energy scale as in Fig. 1.9): the function $\tan \delta_0(E)$ **(a)** compared with $\tan x$ **(b)**

Fig. 1.12 The function $\delta_0(E)$ corresponding to Fig. 1.11a

function decreases to zero with increasing energy. In order to extract the variation of the corresponding phase shift with energy, one should note (Fig. 1.11b), that this behaviour corresponds to the behaviour of the function $\tan x$ in the interval from $x = 0$ to $x = -\pi$. The corresponding energy values are $x = 0 \longrightarrow E = 0$ via $x = -\pi/2 \longrightarrow E = E_S$ to $x = -\pi \longrightarrow E = \infty$. The resolution of the ambiguity of the phase shift by setting $\delta_0(\infty) = 0$ therefore leads to a positive function for the phase shift shown in Fig. 1.12. It reaches the value π for $E = 0$.

3. The pattern indicated continues. If the potential parameters are in the range

$$\bullet \quad \frac{3\pi}{2} \leq [2m_0 v_0]^{1/2} \frac{R}{\hbar} < \frac{5\pi}{2},$$

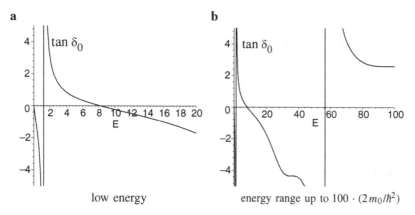

Fig. 1.13 The function $\tan \delta_0$ for S-wave scattering by a potential well with two bound states for low (**a**) and high (**b**) energy values

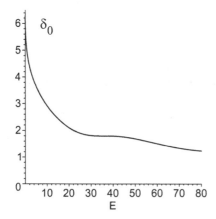

Fig. 1.14 The function $\delta_0(E)$ corresponding to Fig. 1.13

so that two bound states exist, one obtains the behaviour shown in Fig. 1.13 for the function $\tan \delta_0(E)$. The range of this function is $0 \geq \delta_0(E) \geq -2\pi$. The definition of the phase shift for high energies at the value zero requires that a value $\delta_0(0) = 2\pi$ is obtained for very low energies. The variation of the corresponding phase shift with energy is shown in Fig. 1.14.

The behaviour of the phase shifts with energy, explicitly illustrated in these examples, is valid more generally. For short-range attractive potentials the *Theorem of Levinson*[9]

$$\delta_l(0) - \delta_l(\infty) = n_l \pi \tag{1.29}$$

[9] N. Levinson, Danske Videnskab. Selskab, Mat.-fys. Medd. **25**, No 9 (1949).

is valid. It states that, for a given angular momentum quantum number l, the difference of the phase shifts at high and low energies is related to the number n_l of possible states bound in the same potential well. This demonstrates in a different way, that one can obtain information about some aspects of the scattering potential from scattering data.[10]

The limiting case of an impenetrable sphere

$$v(r) \longrightarrow \infty \qquad \text{for} \qquad r \leq R$$

is also of interest. In this case the effective wave number is $K \to i\infty$ and from the formula (1.27) follows for example

$$\tan \delta_0 = -\tan kR \qquad \longrightarrow \qquad \delta_0 = -kR.$$

The corresponding partial cross section

$$\sigma_0 = \frac{4\pi}{k^2} (\sin kR)^2$$

has the limiting value $\sigma_0 \to 4\pi R^2$ for low energies. The increase compared to the geometric cross section of the sphere is due to interference effects of the wave functions.

Other central potentials for which an analytical solution of S-wave scattering can be obtained,[11] are, for example, the exponential potential

$$v(r) = -v_0 e^{-\mu r}$$

and the Hulthen potential

$$v(r) = \frac{v_0 e^{-\mu r}}{1 - e^{\mu r}}.$$

1.3   Scattering by a Coulomb Potential

The discussion of scattering by a Coulomb potential

$$v(r) = \pm \frac{Z_1 Z_2 e^2}{r}$$

---

[10] The proof of Levinson's theorem is discussed in Sect. 5.3.3.

[11] See for example S. Flügge: Practical Quantum Mechanics. SpringerVerlag, Heidelberg (1974).

necessitates a different approach as the boundary condition (1.6) can not be satisfied for a long range potential. The Schrödinger equation

$$\left(-\frac{\hbar^2}{2m_0}\Delta \pm \frac{Z_1 Z_2 e^2}{r} \right) \psi(r) = E\psi(r)$$

is usually written in the form

$$\left(\Delta + k^2 - \frac{2\eta k}{r} \right) \psi(r) = 0. \tag{1.30}$$

In (1.30) the *velocity dependent* Coulomb parameter η

$$\eta = \pm \frac{m_0 Z_1 Z_2 e^2}{\hbar^2 k} \ ,$$

which is also known as the *Sommerfeld parameter*, has been introduced. A different form of this parameter uses the velocity $v = \hbar k/m_0$

$$\eta = \pm \frac{Z_1 Z_2 e^2}{\hbar v}.$$

Two methods are available for solving the Coulomb scattering problem:

- The partial wave expansion can be applied as in Sect. 1.2, however without the requirement of the boundary condition (1.6).
- It is possible to obtain a closed form for the solution, if one uses parabolic coordinates.

1.3.1 Partial Wave Expansion

The ansatz

$$\psi(r) = \sum_{l=0}^{\infty} \frac{w_l(r)}{r} P_l(\cos\theta)$$

leads to a differential equation for the radial part w_l

$$\frac{d^2 w_l(r)}{dr^2} + \left[k^2 - \frac{2\eta k}{r} - \frac{l(l+1)}{r^2} \right] w_l(r) = 0. \tag{1.31}$$

An explicit solution of this differential equation can be obtained with the substitution $x = kr$ and separation of the function $w_l(x)$ into three factors for $x \longrightarrow 0$ (dominated by the centrifugal barrier), for $x \longrightarrow \infty$ (for the case of positive energy) and a remainder F_l

$$w_l(x) = (x)^{l+1} e^{ix} F_l(x).$$

The differential equation for the remainder is

$$x\frac{d^2 F_l(x)}{dx^2} + [2(l+1) + 2ix]\frac{d F_l(x)}{dx} + [2i(l+1) - 2\eta]F_l(x) = 0,$$

which is changed to the simpler form

$$z\frac{d^2 F_l(z)}{dz^2} + [2(l+1) - z]\frac{d F_l(z)}{dz} - [(l+1) + i\eta]F_l(z) = 0$$

by the introduction of the variable $z = -2ix$. This differential equation defines a confluent hypergeometric function $F_l(z) \equiv F(a, c, z)$ with the parameters $a = l+1+i\eta$ and $c = 2l+2$ as well as the variable $z = -2ikr$. Some properties of this function, also called Kummer's function, are compiled in Sect. 1.5.6. As a result of the boundary condition at the coordinate origin, only the regular solutions of the differential equation (1.31), the regular Coulomb functions

$$F_l(\eta, kr) = C_l(kr)^{l+1} e^{ikr} F(l+1+i\eta, 2l+2, -2ikr) \tag{1.32}$$

are admitted. The normalisation factor C_l is usually chosen so that the function $F_l(\eta, kr)$ goes over into the regular solution of the free particle problem $u_l(kr)$ for the boundary condition (1.6) with $\eta = 0$.

Instead of obtaining the solution (1.31) in two steps, one can substitute $z = -2ikr$ and find Whittaker's differential equation[12]

$$\frac{d^2 w_l(z)}{dz^2} + \left[-\frac{1}{4} - \frac{i\eta}{z} - \frac{l(l+1)}{z^2}\right] w_l(z) = 0.$$

The regular Coulomb function is identical (except for a possible normalisation factor) with the regular Whittaker function

$$w_l(z) = z^{l+1} e^{-z/2} F(l+1+i, 2l+2, z).$$

The first step in the discussion of the solution (1.32) is an investigation of the asymptotic behaviour, especially in comparison with results obtained for the

---

[12] Cf. Abramowitz/Stegun, p. 505.

boundary condition (1.6), valid for short-range potentials. If one uses the asymptotic form of Kummer's function given in Sect. 1.5.6 and some simple formulae from the analysis of complex functions, one obtains[13]

$$F_l(\eta, kr) \xrightarrow{kr \to \infty} C_l \frac{e^{\pi \eta / 2} e^{i\gamma_l} \Gamma(2l + 2)}{2^l \Gamma(l + 1 + i\eta)}$$

$$\cdot \sin(kr - \pi l / 2 + \gamma_l - \eta \ln(2kr)). \tag{1.33}$$

The first two terms of the argument of the sine function in (1.33) correspond to the asymptotic form of a plane wave. In addition, a phase shift γ_l and a logarithmic term in the variable kr are found. The logarithmic term is a consequence of the long range of the potential, which leads to a curved orbit of the classical Coulomb trajectory, even in the asymptotic region. For the Coulomb phase shifts γ_l one finds the relation

$$e^{2i\gamma_l} = \frac{\Gamma(l + 1 + i\eta)}{\Gamma(l + 1 - i\eta)} \tag{1.34}$$

or alternatively

$$\gamma_l = \arg \Gamma(l + 1 + i\eta). \tag{1.35}$$

As might be expected the result is $\gamma_l = 0$, if the particles move freely ($\eta = 0$).

As one has $kr \gg \ln(2kr)$ in the asymptotic region, the logarithmic contribution in the result (1.32) can, to a good approximation, be neglected. In this case, the total asymptotic function for the l-th Coulomb partial wave (with (1.32) and the usual normalisation) corresponds exactly to the asymptotic wave function (1.21). Therefore one can write the partial Coulomb scattering amplitude $f_{l,\text{coul}}$ as

$$f_{l,\text{coul}} = \frac{(2l + 1)}{2ik} \left(e^{2i\gamma_l} - 1 \right) \tag{1.36}$$

and the differential Coulomb cross section in the form

$$\frac{d\sigma}{d\Omega}\bigg|_{\text{coul}} = \sum_{l,l'=0}^{\infty} (f_{l',\text{coul}})^* f_{l,\text{coul}} P_l(\cos\theta) P_{l'}(\cos\theta).$$

Figure 1.15 shows, however, that the Coulomb phase shifts increase with the value of the angular momentum. The partial wave expansion converges slowly.[14]

---

[13] This calculation is carried out explicitly in Sect. 1.5.7. The determination of the normalisation constant C_l and the corresponding normalised solution can also be found there.

[14] Note that smaller values of the parameter η correspond to a higher collision energy.

Fig. 1.15 Coulomb phase shifts γ_l as functions of the angular momentum quantum number l for the parameters $\eta = 0.2,\ 1.0,\ 2.0$

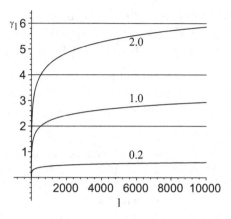

For this reason it is advisable to look for a closed form of the Coulomb wave function $\psi_{\mathrm{coul}}(\mathbf{r})$ and the differential cross section. This is obtained if one solves the Coulomb problem in parabolic coordinates.

1.3.2   Solution in Parabolic Coordinates

Parabolic coordinates can be defined as

$$t = r - z = r(1 - \cos\theta), \quad s = r + z = r(1 + \cos\theta), \quad \varphi = \varphi.$$

The inverse of this transformation is

$$r = \frac{1}{2}(s + t), \qquad z = \frac{1}{2}(s - t).$$

The Schrödinger equation for the Coulomb problem in these coordinates is[15]

$$\left[-\frac{4}{(s+t)} \left\{ \frac{\partial}{\partial t}\left(t\frac{\partial}{\partial t}\right) + \frac{\partial}{\partial s}\left(s\frac{\partial}{\partial s}\right) \right\} \right.$$
$$\left. -\frac{1}{st}\frac{\partial^2}{\partial^2\varphi} + \frac{4\eta k}{(s+t)} \right] \psi(s, t, \varphi) = k^2 \psi(s, t, \varphi).$$

$$(1.37)$$

The separation ansatz $\psi = f_1(s, t) f_2(\varphi)$ shows, that a discussion of the dependence on the angle φ is not necessary in view of the cylindrical symmetry. It is sufficient

---

[15] See P. Moon, D. Eberle: Field Theory Handbook. Springer Verlag, Heidelberg (1961).

to consider the solution with the azimuthal quantum number $m = 0$. The ansatz

$$f_1(s, t) = e^{ikz} f(t) = e^{ik(s-t)/2} f(t)$$

for the remaining function leads to the differential equation[16]

$$t \frac{d^2 f(t)}{dt^2} + (1 - ikt) \frac{df(t)}{dt} - \eta k f(t) = 0.$$

Transition to the variable $x = ikt$ results in

$$x \frac{d^2 f(x)}{dx^2} + (1 - x) \frac{df(x)}{dx} + i\eta f(x) = 0.$$

This is also a differential equation for a Kummer function, this time with the parameters $a = -i\eta$, $c = 1$ and the variable ikt

$$f(t) = C\, F(-i\eta, 1, ikt).$$

The behaviour of the solution

$$f_1(r, z) = C\, e^{ikz} F(-i\eta, 1, ik(r - z)) \tag{1.38}$$

in the asymptotic region can be extracted as before with the asymptotic limit of Kummer's function. This step is executed explicitly in Sect. 1.5.8. The result can be given in the (suggestive!) form

$$f_1(r, z) \xrightarrow{kr \to \infty} \left\{ e^{i(kz + \eta \ln kr(1 - \cos\theta))} \right.$$
$$\left. + \frac{1}{r} \left[-\frac{\eta\, e^{i(2\gamma_0 - \eta \ln(1 - \cos\theta))}}{k(1 - \cos\theta)} \right] e^{i(kr - \eta \ln kr)} \right\}, \tag{1.39}$$

if one chooses

$$C = e^{-\eta\pi/2} \Gamma(1 + i\eta).$$

Compared to the case of a short-range potential, both the incoming plane wave and the outgoing spherical wave are modified by a logarithmic term. This expresses, as in the partial wave form of the scattering solution, the fact that (classical) particles in a Coulomb potential do not move in a straight line, even if they are far from the

---

[16] Do this calculation in order to gain some practice.

source of the potential. The term in the large square bracket can be interpreted as the Coulomb scattering amplitude

$$f_{\text{coul}}(\theta) = \left[-\frac{\eta \exp[2i\gamma_0 - i\eta \ln(1 - \cos\theta)]}{2k \sin^2(\theta/2)} \right], \tag{1.40}$$

since the logarithmic terms in (1.39) only contribute in higher order in $1/r$ to the differential cross section according to (1.7). The differential scattering cross section of the Coulomb problem is

$$\frac{d\sigma}{d\Omega}\bigg|_{\text{coul}} = |f_{\text{coul}}(\theta)|^2 = \frac{\eta^2}{4k^2 \sin^4(\theta/2)}. \tag{1.41}$$

This is the result derived by Rutherford in 1911 on the basis of a *classical* trajectory for a comet. The Rutherford scattering cross section is the only example for which classical mechanics and quantum mechanics give the same result.[17] Two properties of (1.41) can be noticed immediately:

- The differential cross section depends only on the square of the Sommerfeld parameter. It has the same value for attractive or repulsive charges. The scattering of electrons by protons and of positrons by protons leads to the same cross section at the same collision energy.
- The differential cross section diverges for $\theta \to 0$. The reason is apparent if one considers not only the asymptotic limit but the complete asymptotic expansion of Kummer's function. The expansion corresponds to a power series in[18]

$$\frac{1}{ikr(1 - \cos\theta)} \quad \text{with the condition} \quad \left| \frac{1}{kr(1 - \cos\theta)} \right|^2 \ll 1,$$

which can not be satisfied for $\theta = 0$.

It is possible to invoke three options—not pursued here—to avoid this (mathematical) difficulty:

- The quantity kr is very large in actual experiments, so that the corresponding critical angular range (in the forward direction) is so small that it can not be resolved.

---

[17] The motion of planets and comets under the influence of gravitation is the classical counterpart to the discussion of bound and scattering states of a charged particle moving in an electrostatic field of a point charge. See e.g. Dreizler/Lüdde, Vol.1, Sect. 4.1.3.2.

[18] See, e.g. Abramowitz/Stegun, p. 503.

- If a finite charge distribution is considered instead of a point charge, no divergence occurs. An example is the scattering of a (point) electron by a nucleus, which is characterised by a finite charge distribution.
- In an experiment, the target particle is e.g. part of a foil. Therefore it has to be represented by a charge distribution rather than an isolated point charge. Inclusion of the shielding of the target charge prevents the divergence.[19]

1.4   Potential Scattering of Particles with Spin

The discussion begins again by looking at the scattering of two particles, this time with spin s_1 and s_2 (of special interest is $s_1 = s_2 = 1/2$), by a (centrally symmetric) short-range interaction, characterised by asymptotic boundary conditions. The initial state is supposed to represent the scattering of a beam of polarised particles by a polarised target. If one denotes the projection of the spin of the particles onto the beam axis, the z-direction, with μ_1 and μ_2 the initial state at large separation is described by

$$\psi^{in}_{s_1\mu_1,s_2\mu_2}(\boldsymbol{r}) = e^{ikz}\chi_{s_1\mu_1}(1)\chi_{s_2\mu_2}(2). \tag{1.42}$$

The plane wave for the relative motion is multiplied by the spin functions of the two particles. This ansatz is called the μ-*representation* (also the m representation). An equivalent representation of the initial state, useful for many theoretical considerations, is the *channel spin representation*. The factor of the asymptotic wave is in this case the coupled spin wave function χ, which depends on the total spin S, its projection M_S on the z-axis and on the magnitudes of the individual spins of the two particles

$$\tilde{\psi}^{in}_{SM_S,s_1s_2}(\boldsymbol{r}) = e^{ikz}\chi_{SM_S,s_1s_2}(1,2). \tag{1.43}$$

The (Clebsch-Gordan) coupled spin wave function χ_{SM_S,s_1s_2}

$$\chi_{SM_S,s_1s_2}(1,2) =$$

$$\sum_{\mu_1}\begin{bmatrix} s_1 & s_2 & S \\ \mu_1 & M_S - \mu_1 & M_S \end{bmatrix}\chi_{s_1\mu_1}(1)\chi_{s_2M_S-\mu_1}(2) \tag{1.44}$$

---

[19] This point is addressed, for example, in the book by J. R. Taylor: Scattering Theory. John Wiley, New York (1972), Chap. 14-a.

is used here.[20] The channel spin representation has the advantage that the spin function possesses a defined permutation symmetry in the case of two identical particles with $s_1 = s_2$. This simplifies the discussion of questions concerning spin and statistics.

The ansatz for the scattering wave function must take the possibility into account that the spin orientation can change due to the interaction. For this reason, one writes in the μ-representation

$$\psi^{\text{scat}}_{s_1\mu_1,s_2\mu_2}(r) \xrightarrow{r\to\infty} \sum_{\mu'_1,\mu'_2} f^{s_1,s_2}_{\mu_1\mu_2,\mu'_1\mu'_2}(\theta,\varphi) \frac{e^{ikr}}{r} \chi_{s_1\mu'_1}(1)\chi_{s_2\mu'_2}(2). \tag{1.45}$$

The magnitude of the spin—an intrinsic property of the particles—does not change. However, a statistically weighted transition to every combination of spin orientations is possible, e.g. for half integer spin $s_1 = s_2 = 1/2$

$$\uparrow\uparrow \quad\longrightarrow\quad \uparrow\uparrow, \quad \uparrow\downarrow, \quad \downarrow\uparrow, \quad \downarrow\downarrow.$$

In this example there exist four spin-indexed scattering amplitudes (that is a total of 16 combinations for the four possible initial channels).

In the channel spin representation, the ansatz for the asymptotic scattering state is

$$\tilde{\psi}^{\text{scat}}_{SM_S,s_1s_2}(r) \xrightarrow{r\to\infty} \sum_{M'_S} \tilde{f}^{S,s_1s_2}_{M_S,M'_S}(\theta,\varphi) \frac{e^{ikr}}{r} \chi_{SM'_S,s_1s_2}(1,2). \tag{1.46}$$

A change of the channel spin $S \to S'$ does not occur for the standard interactions, so that one is dealing with a more economic form. The relation between the scattering amplitudes in the two representations follows from the properties of the Clebsch-Gordan coefficients as

$$f^{s_1,s_2}_{\mu_1\mu_2,\mu'_1\mu'_2} = \sum_S \begin{bmatrix} s_1 & s_2 & S \\ \mu_1 & \mu_2 & M_S \end{bmatrix} \begin{bmatrix} s_1 & s_2 & S \\ \mu'_1 & \mu'_2 & M'_S \end{bmatrix} \tilde{f}^{S,s_1,s_2}_{M_S,M'_S} \tag{1.47}$$

and a corresponding expression for the inverse transformation.

---

[20] For the coupling of two states with angular momenta (l_1, m_1) and (l_2, m_2) to a state with total angular momentum (L, M) (and correspondingly of l with s to j, s_1 with s_2 to S and j_1 with j_2 to J) one uses the ansatz

$$\Psi_{L,M,l_1,l_2}(\Omega_1,\Omega_2) = \sum_{m_1} \begin{bmatrix} l_1 & l_2 & L \\ m_1 & M-m_1 & M \end{bmatrix} \psi_{l_1,m_1}(\Omega_1)\psi_{l_2,M-m_1}(\Omega_2)$$

and determines the expansion coefficients by demanding that the state characterised by the wave function Ψ should be a state of the total angular momentum (L, M). Further information can be found e.g. in M. E. Rose, Elementary Theory of Angular Momentum. Dover Publications, New York (1995).

1.4.1 Differential Cross Sections

The actual quantity observed is the differential cross section, for which—depending on the analysis of the spin projections with respect to the direction of the incident beam—the following possibilities are possible:

- If the spin orientations of the two particles are registered before and after the collision, one finds in extension of the definition for spinless particles[21]

$$\frac{d\sigma}{d\Omega}\Big|_{\mu_1\mu_2\to\mu_1'\mu_2'} = \left| f^{s_1,s_2}_{\mu_1\mu_2,\mu_1'\mu_2'}(\theta,\varphi) \right|^2. \tag{1.48}$$

- If one does not analyse the spin projections in the final state, but only observes that the particles arrive in the detector, one measures

$$\frac{d\sigma}{d\Omega}\Big|_{\mu_1\mu_2\to\text{all}} = \sum_{\mu_1'\mu_2'} \left| f^{s_1,s_2}_{\mu_1\mu_2,\mu_1'\mu_2'}(\theta,\varphi) \right|^2.$$

- If an unpolarised particle of the beam scatters from an unpolarised particle in the target, then, under the assumption that all initial orientations of the spin are equally likely, one averages over the initial orientations. The spin projections in the final state might or might not be registered. In the latter case, the measured quantity is

$$\frac{d\sigma}{d\Omega}\Big|_{\text{unpol}} = \frac{1}{(2s_1+1)}\frac{1}{(2s_2+1)} \sum_{\mu_1\mu_2\mu_1'\mu_2'} \left| f^{s_1,s_2}_{\mu_1\mu_2,\mu_1'\mu_2'}(\theta,\varphi) \right|^2.$$

If one substitutes the transformation between the two representations of the scattering amplitude on the right-hand side, one finds

$$\frac{d\sigma}{d\Omega}\Big|_{\text{unpol}} = \frac{1}{(2s_1+1)}\frac{1}{(2s_2+1)} \sum_{SM_SM_S'} \left| \tilde{f}^{S,s_1s_2}_{M_S,M_S'}(\theta,\varphi) \right|^2.$$

The unpolarised differential cross section can be calculated directly with the aid of the channel spin scattering amplitude.
- Other combinations, such as a beam of particles with certain initial spin projection colliding with an unpolarised target, are possible.

---

[21] It is assumed that the spin orientation (polarisation) is 100% in each case. This is not achievable in practice. Partial polarisation (e.g. a beam with 80% spin up and 20% spin down) can be discussed in terms of the spin density formalism, see Chap. 5.

1.4.2 Selection Rules

The interaction determines the details of possible transitions into final channels, respectively the form of the corresponding partial wave expansion. For a central potential

$$v(r) \longrightarrow \hat{V}_\mathrm{c}$$

the commutation relations

$$[\hat{V}_\mathrm{c}, \hat{l}] = \mathbf{0} \qquad \text{and} \qquad [\hat{V}_\mathrm{c}, \hat{S}] = \mathbf{0}$$

are valid. The components of the relative angular momentum of the two particles are denoted by $\hat{l}$ and those of the total spin by $\hat{S} = \hat{s_1} + \hat{s_2}$. The commutation relations imply, that neither the angular momentum quantum numbers l and m nor the spin quantum numbers S and M_S can change during the scattering by a central potential. The potential does not affect the spin of the particles, so that

$$\tilde{f}^{S,s_1 s_2}_{M_S, M'_S}(\theta, \varphi) = \delta_{M_S, M'_S} f(\theta, \varphi)$$

follows, as well as immediately with (1.47)

$$f^{s_1, s_2}_{\mu_1 \mu_2, \mu'_1 \mu'_2}(\theta, \varphi) = \delta_{\mu_1, \mu'_1} \delta_{\mu_2, \mu'_2} f(\theta, \varphi).$$

There is de facto no difference compared with the scattering of spinless particles. The spin part of the wave function is the same in the initial and in the final state (and does not change during the whole scattering process).

For a central potential with spin-spin interaction, which plays e.g. a role in nuclear physics,

$$v(r) \longrightarrow \hat{V}_\mathrm{ss} \longrightarrow v(r) + v_\mathrm{s} \cdot (\hat{s_1} \cdot \hat{s_2})$$

one still has

$$[\hat{V}_\mathrm{ss}, \hat{l}] = \mathbf{0} \qquad \text{and} \qquad [\hat{V}_\mathrm{ss}, \hat{S}] = \mathbf{0},$$

so that the selection rules are the same as in the case of a pure central potential. However, since the expectation value of the spin-spin operator has the value

$$\langle SM_S, s_1 s_2 | \hat{s_1} \cdot \hat{s_2} | SM_S, s_1 s_2 \rangle = \frac{\hbar^2}{2} \left(S(S+1) - s_1(s_1+1) - s_2(s_2+1) \right)$$

the scattering amplitude must depend on the quantum number S

$$\tilde{f}^{S,s_1 s_2}_{M_S, M'_S}(\theta, \varphi) = \delta_{M_S, M'_S} f^S(\theta, \varphi).$$

This means, for example, that the singlet ($S = 0$) and the triplet scattering ($S = 1$) of spin-1/2 particles are different. The partial scattering amplitudes and the phase shifts have to be distinguished by the spin quantum number S

$$f^S_l = \frac{(2l + 1)}{k} \, e^{i\delta_{lS}(k)} \sin \delta_{lS}(k).$$

The situation is much more complicated for scattering by a spin-orbit potential of the form[22]

$$v(r) \longrightarrow \hat{V}_{\text{sl}} \longrightarrow v(r) + v_{\text{sl}} \cdot (\hat{S} \cdot \hat{l}).$$

The commutation relations are in this case

$$[\hat{V}_{\text{sl}}, \hat{l}] \neq 0, \qquad [\hat{V}_{\text{sl}}, \hat{S}] \neq 0,$$

but on the other hand one also finds

$$[\hat{V}_{\text{sl}}, \hat{l}^2] = [\hat{V}_{\text{sl}}, \hat{S}^2] = [\hat{V}_{\text{sl}}, \hat{J}^2] = [\hat{V}_{\text{sl}}, \hat{J}_z] = 0,$$

where $\hat{J} = \hat{l} + \hat{S}$ is the total angular momentum of the two-particle system. In this case, the partial wave expansion must be applied in a different form, namely as an expansion in terms of eigenfunctions of the total angular momentum and its projection onto the beam axis. For example, in the simple case with $s_1 = 1/2$ and $s_2 = 0$, scattering with spin flip can occur

$$\left.\begin{array}{c}(\uparrow, 0) \\ (\downarrow, 0)\end{array}\right\} \longrightarrow \quad a^{(+)}(\uparrow, 0) \quad + \quad a^{(-)}(\downarrow, 0).$$

1.4.3 Role of Particle Statistics

The Pauli principle plays a role in the scattering of identical particles. The total wave function must be symmetric for two bosons or antisymmetric for two fermions. If one deals with $s = 0$ bosons or $s = 1/2$ fermions, one must consider the following possibilities:

---

[22] In the case of two particles scattering from each other, the total spin and the relative orbital angular momentum are to be used here.

- A symmetric function must be used for the space part in the case of two bosons or two fermions with total spin 0.
- An antisymmetric function must be used for the space part in the case of two fermions with total spin 1.

The space part can be expressed in the form

$$\psi_{\text{sym/anti}}(\boldsymbol{r}) = \frac{1}{\sqrt{2}}\left[1 \pm \hat{P}_{12}\right]\psi(\boldsymbol{r}). \tag{1.49}$$

The permutation operator $\hat{P}_{12}$ causes a change of sign of the relative coordinate

$$\boldsymbol{r} \Longrightarrow -\boldsymbol{r} \quad \text{or} \quad r, \theta, \varphi \Longrightarrow r, \pi - \theta, \varphi + \pi.$$

The differential cross section obtained from the asymptotic form of a symmetric or an antisymmetric wave function is, respectively,

$$\left.\frac{d\sigma}{d\Omega}\right|_{\text{sym/anti}} = \left|f(\theta, \varphi) \pm f(\pi - \theta, \varphi + \pi)\right|^2. \tag{1.50}$$

The particle statistics leads to a typical quantum mechanical interference of the two scattering amplitudes.

As examples, one may consider the following situations:

- The differential cross section for the scattering of two identical particles with spin s_1 in an experiment without measurement of the polarisation is generally

$$\left.\frac{d\sigma}{d\Omega}\right|_{\text{unpol}} = \frac{1}{(2s_1 + 1)^2}\sum_{SM_S}\left|f^S(\theta) + (-1)^S f^S(\pi - \theta)\right|^2$$

$$= \frac{1}{(2s_1 + 1)^2}\sum_{S}(2S + 1)\left|f^S(\theta) + (-1)^S f^S(\pi - \theta)\right|^2.$$

In particular, for the case of two spin-1/2 fermions, one finds

$$\left.\frac{d\sigma}{d\Omega}\right|_{\text{unpol}} = \frac{1}{4}\left|f^0(\theta) + f^0(\pi - \theta)\right|^2 + \frac{3}{4}\left|f^1(\theta) - f^1(\pi - \theta)\right|^2.$$

- For the scattering of two particles with an interaction that depends only on the distance between the particles, the scattering amplitude is

$$f(\theta) = \sum_l f_l(k)P_l(\cos\theta).$$

With the relations

$$\cos(\pi - \theta) = -\cos\theta \quad \text{and} \quad P_l(-\cos\theta) = (-1)^l P_l(\cos\theta)$$

one arrives at the statement

$$\frac{d\sigma}{d\Omega}\bigg|_{s/a} = \sum_{ll'} \left(1 \pm (-1)^l\right)\left(1 \pm (-1)^{l'}\right) f_l^*(k) f_{l'}(k) P_l(\cos\theta) P_{l'}(\cos\theta).$$

Even l- and l'-values contribute for symmetric space parts, odd values for anti-symmetric space parts. This results in markedly different angular distributions.

• In the case of Coulomb scattering, one obtains from the result (1.40) a symmetric and an antisymmetric form of the differential cross section

$$\frac{d\sigma}{d\Omega}\bigg|_{s/a} = \frac{\eta^2}{4k^2} \left\{ \frac{1}{\sin^4\frac{\theta}{2}} + \frac{1}{\cos^4\frac{\theta}{2}} \pm \frac{8}{\sin^2\theta} \cos(\eta \ln \tan^2\theta/2) \right\}, \tag{1.51}$$

which are known under the name of *Mott cross section*. This cross section contains two classical and one interference terms. Concerning the interference (or exchange) term one can note the following:

– If the Sommerfeld parameter is large (the relative energy small), then the interference term is a rapidly oscillating function of the angle θ, which in general can not be resolved in the experiment (Fig. 1.16a, b). One measures in this case

$$\frac{d\sigma}{d\Omega}\bigg|_{s/a} \xrightarrow{k\to 0} \frac{\eta^2}{4k^2} \left\{ \frac{1}{\sin^4\theta} + \frac{1}{\cos^4\theta} \right\}.$$

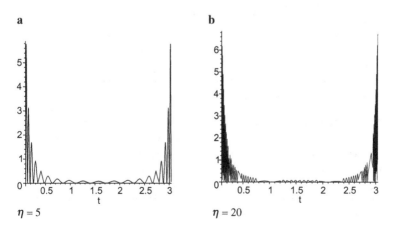

a

$\eta = 5$

b

$\eta = 20$

Fig. 1.16 Interference term in the Mott cross section in the range $0.1 \le t = \cos\theta \le \pi - 0.1$ (in units of k^2/η^2, scaled by 10^{-2}) for two values of the Sommerfeld parameter η

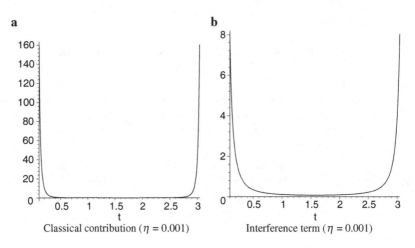

Fig. 1.17 The classical part **(a)** and the interference term **(b)** of the Mott cross section in the range $0.2 \leq t = \cos\theta \leq \pi - 0.2$ for $\eta = 0.001$ (in units of k^2/η^2 scaled by 10^{-2})

This is quite a classical result in the form of the sum of two intensities as shown in Fig. 1.17a for a lower η-value.

- At high collision energies (small Sommerfeld parameter) the interference term is a smooth function of $t = \cos\theta$, as illustrated in Fig. 1.17b. However, it is much smaller than the classical cross section, as can be gleaned from Fig. 1.17a.

1.5   Detailed Calculations for Chap. 1

1.5.1   The Differential Cross Section

In order to calculate the current density

$$j = -\frac{i\hbar}{2\,m_0}\left(\psi^*(r)\left[\nabla\psi(r)\right] - \left[\nabla\psi^*(r)\right]\psi(r)\right)$$

one needs the gradients

$$\nabla\psi_{\text{in}}(r) = ik e^{ikz} e_z,$$

$$\nabla\psi_{\text{scat}}(r) = \left\{\left(\frac{ik}{r} - \frac{1}{r^2}\right) f e^{ikr} e_r + \left(\frac{1}{\sin\theta}\frac{\partial f}{\partial\varphi} e_\varphi + \frac{\partial f}{\partial\theta} e_\theta\right)\frac{e^{ikr}}{r^2}\right\}$$

evaluated with the asymptotic scattering solution

$$\psi(r) \longrightarrow e^{ikz} + f(\theta, \varphi)\frac{e^{ikr}}{r}.$$

In the asymptotic region one can restrict oneself to the leading order in $1/r$ and thus neglect all contributions in $1/r^2$. For the magnitudes of the current densities one obtains therefore

$$j_{\text{in}}(r) = \frac{\hbar k}{m_0}, \qquad j_{\text{scat}}(r) = \frac{\hbar k}{m_0}\frac{f^* f}{r^2},$$

so that the result for the differential cross section is

$$\frac{d\sigma}{d\Omega} = \frac{r^2 j_{\text{scat}}}{j_{\text{in}}} = |f(\theta, \varphi)|^2.$$

1.5.2   The Optical Theorem

The scalar product of the asymptotic limit of the current density vector with the vector of an infinitesimal area on a sphere is

$$j(r) \cdot df = -\frac{i\hbar}{2m_0}\Big(\psi^*(r)[\partial_r \psi(r)] - [\partial_r \psi^*(r)]\psi(r)\Big)r^2 d\Omega.$$

The partial derivatives of the asymptotic wave function with respect to the radial coordinate are taken from Sect. 1.5.1. Restriction to leading order terms in $1/r$ and use of $x = \cos\theta$, leads to the integral

$$\iint j \cdot df = \frac{\hbar k}{m_0}\lim_{r\to\infty}\iint dx\, d\varphi\,\Big\{r^2 x + r(1+x)\text{Re}\,[f]\cos(kr(1-x))$$
$$-\,\text{Im}\,[f]\sin(kr(1-x)) + |f|^2\Big\}.$$

The different terms correspond to the contribution of the plane wave, the interference terms between the two parts of the wave function and the contribution of the scattered wave. The result for the first and the last term can be given directly. It is

$$\int d\varphi \int_{-1}^{1} dx\, x = 0,$$

$$\iint |f(\theta, \varphi)|^2 d\Omega = \sigma.$$

For the evaluation of the interference terms one considers the integrals

$$I_1(r, \varphi) = \int_{-1}^{1} dx \, (1+x) F_1(x, \varphi) \cos[kr(1-x)],$$

$$I_2(r, \varphi) = \int_{-1}^{1} dx \, (1+x) F_2(x, \varphi) \sin[kr(1-x)].$$

With the substitution $y = 1 - x$ and subsequent partial integration one obtains

$$I_1(r, \varphi) = \int_{0}^{2} dy \, (2-y) F_1(1-y, \varphi) \cos[kry]$$

$$= 0 - \frac{1}{kr} \int_{0}^{2} dy \, \frac{\partial[(2-y)F_1(1-y, \varphi)]}{\partial y} \sin[kry],$$

$$I_2(r, \varphi) = \int_{0}^{2} dy \, (2-y) F_2(1-y, \varphi) \sin[kry]$$

$$= \frac{2}{kr} F_2(1, \varphi) + \frac{1}{kr} \int_{0}^{2} dy \, \frac{\partial[(2-y)F_2(1-y, \varphi)]}{\partial y} \cos[kry].$$

Since each additional partial integration generates a higher power in $1/r$, one can neglect the remaining integrals in the asymptotic region, and write

$$I_1(r, \varphi) \to 0 + \mathcal{O}\left(\frac{1}{r^2}\right) \quad \text{and} \quad I_2(r, \varphi) \to \frac{2}{kr} F_2(1, \varphi) + \mathcal{O}\left(\frac{1}{r^2}\right).$$

With these results one obtains the optical theorem in the form

$$\frac{\hbar k}{m_0} \left\{ \frac{2}{k} \int d\varphi \, \mathrm{Im} f(0, \varphi) - \iint d\Omega \, |f(\theta, \varphi)|^2 \right\} = 0$$

or, as the scattering amplitude in the forward direction does not depend on the angle φ

$$\frac{4\pi}{k} \mathrm{Im} f(0, \varphi) - \sigma \equiv \frac{4\pi}{k} \mathrm{Im} f(0) - \sigma = 0.$$

1.5.3   Centre of Mass and Laboratory Systems

The positions of the collision partners (with masses m_{10} and m_{20}) from the point of view of the laboratory system (characterised by lower case letters) and from the point of view of the centre of mass system (characterised by upper case letters) are shown in Fig. 1.18. The position of the centre of mass is (use r_{cm} instead of the

Fig. 1.18 Position vectors

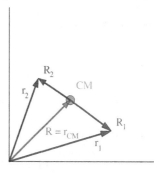

usual R. M is the total mass $M = m_{10} + m_{20}$)

$$r_{cm} = \frac{1}{M}(m_{10}r_1 + m_{20}r_2).$$

No forces act on the centre of mass. The momentum of the centre of mass is therefore a conserved quantity.

$$P_{cm} = p_1 + p_2 = p'_1 + p'_2 = p'_{cm}.$$

The primed quantities refer to the situation after the collision. Elastic scattering is characterised by conservation of kinetic energy (outside the finite range in which the interaction is present)

$$E = \frac{p_1^2}{2m_{10}} + \frac{p_2^2}{2m_{20}} = \frac{p'^2_1}{2m_{10}} + \frac{p'^2_2}{2m_{20}} = E'. \tag{1.52}$$

The centre of mass system is defined by the statement

$$P_{cm} = 0.$$

The relation between the coordinates and momenta in the two systems is ($i = 1, 2$)

$$R_i = r_i - r_{cm},$$
$$P_i = p_i - \frac{m_{i0}}{M}P_{cm}. \tag{1.53}$$

Conservation of the momentum of the centre of mass involves, from the point of view of the centre of mass system

$$P_1 + P_2 = P'_1 + P'_2 = 0. \tag{1.54}$$

Fig. 1.19 Particle momentum in the centre of mass system, (**a**) before the collision and (**b**) after the collision

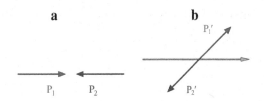

The momentum vectors of the two particles, before (unprimed quantities) as well as after the collision (primed quantities), have an opposite direction (Fig. 1.19).

Rewriting the total kinetic energy of the system (1.52) with the transformation (1.53) results in

$$E = \frac{p_{cm}^2}{2M} + \frac{P_1^2}{2m_{10}} + \frac{P_2^2}{2m_{20}}.$$

In the centre of mass system, only the sum of the last two terms, the kinetic energy of the relative motion, is observed. As a result of the conservation of momentum (1.54) one therefore finds

$$E_{\text{rel}} = \frac{P_1^2}{2\mu} = E'_{\text{rel}} = \frac{P_1'^2}{2\mu}.$$

The quantity $\mu = m_{10} m_{20}/M$ is the reduced mass.

The magnitude of the momentum vectors before and after the collision is unchanged. For the relative momentum before the collision

$$p_{\text{rel}} = \mu(\dot{r}_1 - \dot{r}_2)$$

one obtains by rewriting (1.53) and using (1.54)

$$p_{\text{rel}} = P_{\text{rel}} = \frac{(m_{20}P_1 - m_{10}P_2)}{M} = P_1.$$

After the collision one finds for the relative momentum from the point of view of the centre of mass

$$P'_{\text{rel}} = P'_1 \neq P_1.$$

The relative momentum can change its direction, but not its magnitude.

The scattering angle θ_{cm} in the centre of mass system is the angle between the momentum vectors of particle 1 (or the relative momentum vectors) before and after the collision

$$\cos \theta_{\text{cm}} = \frac{P'_1 \cdot P_1}{P_1^2},$$

Fig. 1.20 Relation between
the scattering angles

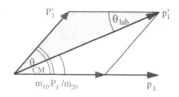

the differential cross-section from the point of view of the centre of mass system is

$$\left(\frac{d\sigma}{d\Omega}\right)_{cm} = \left(\frac{d\sigma}{d\cos\theta_{cm}d\varphi}\right)_{cm}.$$

In the experiment, which takes place in the laboratory system, the angle between the vectors p_1 and p'_1 is measured

$$\cos\theta_{lab} = \frac{p_1 \cdot p'_1}{|p_1||p'_1|}.$$

For comparison with the centre of mass system, a conversion is necessary. This conversion is most simple in the case that the target particle (m_{20}) is at rest before the collision[23]

$$p_2 = 0.$$

With this assumption one can note the relations

$$P_{cm} = p_1 + p_2 \qquad \longrightarrow \qquad P_{cm} = p_1,$$

$$p_1 = P_1 + \frac{m_{10}}{M}P_{cm} \qquad \longrightarrow \qquad p_1 = \frac{M}{m_{20}}P_1,$$

$$p'_1 = P'_1 + \frac{m_{10}}{M}P_{cm} \qquad \longrightarrow \qquad p'_1 = \frac{m_{10}}{m_{20}}P_1 + P'_1.$$

In Fig. 1.20 the momentum vectors of particle 1 and the scattering angles from the point of view of the two reference frames are illustrated for this special case. The vectors p_1 and P_1 are colinear. The difference of the vectors p'_1 and P'_1 corresponds to a vector which is colinear with p_1, respectively with P_1

$$P'_1 - p'_1 = \frac{m_{10}}{m_{20}}P_1 = \frac{m_{10}}{M}p_1.$$

---

[23] The calculation for the case of a head on collision of two particle beams would be a useful exercise.

According to the marked triangle in this figure one finds with the sine theorem the statement

$$\frac{\sin(\theta_{cm} - \theta_{lab})}{\sin \theta_{lab}} = \frac{m_{10} P_1}{m_{20} P'_1} = \frac{m_{10}}{m_{20}}.$$

Resolution leads to a relation between the scattering angles

$$\tan \theta_{lab} = \frac{\sin \theta_{cm}}{m_{10}/m_{20} + \cos \theta_{cm}}$$

respectively

$$\cos \theta_{lab} = \frac{\dfrac{m_{10}}{m_{20}} + \cos \theta_{cm}}{\left[\left(\dfrac{m_{10}}{m_{20}}\right)^2 + 2\dfrac{m_{10}}{m_{20}} \cos \theta_{cm} + 1\right]^{1/2}}.$$

The differential cross section in the laboratory system is therefore connected to the corresponding cross section in the centre of mass system by the relation

$$\left(\frac{d\sigma}{d\Omega}\right)_{lab} = \left(\frac{d\sigma}{d\Omega}\right)_{cm} \left(\frac{d\Omega_{cm}}{d\Omega_{lab}}\right) = \left(\frac{d\sigma}{d\Omega}\right)_{cm} \left|\left(\frac{d\cos\theta_{cm}}{d\cos\theta_{lab}}\right)\right|$$

$$= \left(\frac{d\sigma}{d\Omega}\right)_{cm} \frac{\left[1 + 2(m_{10}/m_{20})\cos\theta_{cm} + (m_{10}/m_{20})^2\right]^{3/2}}{|1 + (m_{10}/m_{20})\cos\theta_{cm}|}. \tag{1.55}$$

1.5.4   Various Bessel Functions

Properties of the Bessel-Riccati functions are:

- Differential equation ($x = kr$, R' means derivative with respect to x)

$$R''_l(x) + \left[1 - \frac{l(l+1)}{x^2}\right] R_l(x) = 0.$$

- Recursion formulae, for example

$$R_{l+1}(x) = \frac{(l+1)}{x} R_l(x) - R'_l(x).$$

- Functions (regular u_l, irregular v_l) with $l = 0, 1, 2$

$$u_0(x) = \sin x, \ v_0(x) = -\cos x,$$

$$u_1(x) = \frac{1}{x}\sin x - \cos x, \ v_1(x) = -\frac{1}{x}\cos x - \sin x,$$

$$u_2(x) = \left(\frac{3}{x^2} - 1\right)\sin x - \frac{3}{x}\cos x, \ v_2(x)\left(1 - \frac{3}{x^2}\right)\cos x - \frac{3}{x}\sin x.$$

- Behaviour for $x \longrightarrow 0$
  The regular solution remains finite at the coordinate origin

$$u_l(x) \longrightarrow \frac{x^{l+1}}{(2l+1)!!},$$

the singular solution diverges

$$v_l(x) \longrightarrow \frac{(2l-1)!!}{x^l}.$$

- Behaviour for $x \longrightarrow \infty$

$$u_l(x) \longrightarrow \sin\left(x - l\frac{\pi}{2}\right),$$

$$v_l(x) \longrightarrow -\cos\left(x - l\frac{\pi}{2}\right).$$

For the spherical Bessel functions $j_l(x), n_l(x)$ one has:

- These functions are related to the Bessel-Riccati functions in a simple way

$$\text{spherical Bessel function} \quad j_l(x) = \frac{u_l(x)}{x},$$

$$\text{spherical Neumann function} \quad n_l(x) = \frac{v_l(x)}{x},$$

so that the properties of these functions can be given directly.
- Differential equation,

$$f_l''(x) + \frac{2}{x}f_l'(x) + \left[1 - \frac{l(l+1)}{x^2}\right]f_l(x) = 0.$$

- Recursion formulae, e.g.

$$f_l'(x) = \frac{l}{x} f_l(x) - f_{l+1}(x).$$

1.5.5 Partial Waves

The matching condition

$$\frac{1}{R_l^{G1}(r)} \frac{dR_l^{G1}(r)}{dr}\bigg|_{r_s} = \frac{1}{R_l^{G2}(r)} \frac{dR_l^{G2}(r)}{dr}\bigg|_{r_s}$$

for a regular Bessel function $R_l(r) = r u_l(r)$ has the form

$$\frac{1}{u_l^{G1}(r)} \frac{du_l^{G1}(r)}{dr}\bigg|_{r_s} = \frac{1}{u_l^{G2}(r)} \frac{du_l^{G2}(r)}{dr}\bigg|_{r_s}.$$

For the matching of the functions in the asymptotic region

$$u_l^{\text{asymp}}(r) = \frac{i^l(2l+1)}{k} \sin(kr - l\pi/2) + f_l e^{ikr}$$

and the intermediate region

$$u_l^I(r) = \frac{N_l}{k} \sin(kr - l\pi/2 + \delta_l)$$

one has to evaluate the condition ($r_s \to r$)

$$\frac{k\cos(kr - l\pi/2 + \delta_l)}{\sin(kr - l\pi/2 + \delta_l)} = \frac{k(i^l(2l+1)\cos(kr - l\pi/2) + ikf_l e^{ikr})}{(i^l(2l+1)\sin(kr - l\pi/2) + kf_l e^{ikr})}.$$

Resolution with respect to f_l yields initially

$$kf_l e^{ikr} \cdot \quad (\cos(kr - l\pi/2 + \delta_l) - i\sin(kr - l\pi/2 + \delta_l))$$
$$= \quad i^l(2l+1)[\cos(kr - l\pi/2)\sin(kr - l\pi/2 + \delta_l)$$
$$- \quad \sin(kr - l\pi/2)\cos(kr - l\pi/2 + \delta_l)].$$

The trigonometric functions on the two sides of this equation can be combined

$$\cos(kr - l\pi/2)\sin(kr - l\pi/2 + \delta_l) - \sin(kr - l\pi/2)\cos(kr - l\pi/2 + \delta_l)$$
$$= \sin\delta_l$$

and

$$\cos(kr - l\pi/2 + \delta_l) - i\sin(kr - l\pi/2 + \delta_l) = e^{-i(kr - l\pi/2 + \delta_l)}$$
$$= i^l e^{-ikr} e^{-i\delta_l}.$$

A relation between the partial scattering amplitude and the corresponding phase shifts follows

$$f_l = \frac{(2l+1)}{k}\sin\delta_l e^{i\delta_l} = \frac{(2l+1)}{2ik}\left(e^{2i\delta_l} - 1\right).$$

Alternatively, one can consider the matching of the wave functions themselves

$$\frac{N_l}{k}\sin(kr - l\pi/2 + \delta_l) = \frac{i^l(2l+1)}{k}\sin(kr - l\pi/2) + f_l e^{ikr}.$$

Using

$$\sin(kr - l\pi/2) = \frac{1}{2i}\left((-i)^l e^{ikr} - (i)^l e^{-ikr}\right)$$

and

$$\sin(kr - l\pi/2 + \delta_l) = \frac{1}{2i}\left((-i)^l e^{i\delta_l} e^{ikr} - (i)^l e^{-i\delta_l} e^{-ikr}\right)$$

one can sort this condition in the form

$$e^{ikr}\left(f_l - \frac{1}{2i}\left[\frac{(2l+1)}{k} - (-i)^l e^{i\delta_l}\frac{N_l}{k}\right]\right)$$
$$+ e^{-ikr}\left(\frac{(i)^{l-1}}{2}\left[\frac{N_l}{k}e^{-i\delta_l} - \frac{(2l+1)}{k}(i)^l\right]\right) = 0.$$

Since the two functions $e^{\pm ikr}$ are linearly independent, the respective coefficients must vanish. For this reason, one extracts the following statements

$$N_l = (2l+1)i^l e^{i\delta_l},$$
$$f_l = \frac{(2l+1)}{2ik}\left[e^{2i\delta_l} - 1\right].$$

The expression for the partial scattering amplitude agrees with the previous result.

1.5.6   The Confluent Hypergeometric Function

The confluent hypergeometric series, also called Kummer's function (with a complex, arbitrary variable z),

$$F(a, c, z) = \frac{\Gamma(c)}{\Gamma(a)} \sum_{0}^{\infty} \frac{\Gamma(a+n)}{\Gamma(c+n)} \frac{z^n}{n!} = 1 + \frac{a}{c} z + \frac{a(a+1)}{c(c+1)} \frac{z^2}{2!} + \ldots \quad (1.56)$$

is the solution of the differential equation, which is regular for $z \longrightarrow 0$

$$zy''(z) + (c - z)y'(z) - ay(z) = 0. \quad (1.57)$$

A linearly independent solution of Kummer's differential equation is

$$y_2(z) = z^{1-c} F(a - c + 1, 2 - c, z),$$

provided c is not a positive integer.

From the long list of properties of the function $F(a, c, z)$ (derivatives, integrals, recursion formulae, integral representations) only three statements are needed here.

- The asymptotic expansion is obtained by evaluating the representation of the function in terms of the complex Barnes contour integral in the limiting case $|z| \longrightarrow \infty$ as

$$\lim_{|z| \to \infty} F(a, c, z) \longrightarrow \frac{\Gamma(c)}{\Gamma(c-a)} e^{i\pi a \epsilon} z^{-a} + \frac{\Gamma(c)}{\Gamma(a)} e^z (z)^{a-c}. \quad (1.58)$$

The parameter ϵ has the value 1 or -1 if the complex number z tends towards ∞ in the upper or the lower half plane. The more precise statement is

$$\epsilon = 1 \qquad \text{if} \qquad 0 < \arg z < \pi,$$

$$\epsilon = -1 \qquad \text{if} \qquad -\pi < \arg z \leq 0.$$

- The Bessel function $j_l(x)$ is a special case of Kummer's function

$$j_l(x) = x^l e^{-ix} \frac{\Gamma(1/2)}{2^{l+1} \Gamma(l + 3/2)} F(l + 1, 2l + 2, 2ix). \quad (1.59)$$

- Useful is also Kummer's theorem

$$e^{-z/2} F(a, c, z) = e^{z/2} F(c - a, c, -z). \quad (1.60)$$

Further information concerning $F(a, c, z)$ can be found in Abramowitz, Stegun.

1.5.7 Properties of the Coulomb Wave Function

Asymptotic Form From the asymptotic limit (1.58) of Kummer's function one obtains the asymptotic form of the Coulomb function

$$F(\eta, kr) = C_l \, (kr)^{l+1} e^{ikr} F(l + 1 + i\eta, 2l + 2, -2ikr)$$

initially as

$$F(\eta, kr) \longrightarrow C_l \, (kr)^{l+1} e^{ikr} \left\{ \frac{\Gamma(2l + 2)}{\Gamma(l + 1 - i\eta)} e^{-i\pi(l+1+i\eta)} (-2ikr)^{-(l+1+i\eta)} \right.$$

$$\left. + \frac{\Gamma(2l + 2)}{\Gamma(l + 1 + i\eta)} e^{-2ikr} (-2ikr)^{-(l+1-i\eta)} \right\}.$$

Since the argument of Kummer's function is $(-2ikr)$, the limiting process takes place in the lower complex half-plane. For the transformation of this expression, one uses

$$i = e^{i\pi/2},$$

in order to write

$$(-2ikr)^{-(l+1\pm i\eta)} = [(-i)(2kr)]^{-(l+1\pm i\eta)}$$

$$= (2kr)^{-(l+1)} e^{\mp i\eta \ln(2kr)} e^{i\pi(l+1)/2} e^{\mp \pi\eta/2}$$

and extracts a part of the terms in the curly brackets. The intermediate result is

$$F(\eta, kr) \longrightarrow C_l \, \frac{e^{\pi\eta/2}}{2^{l+1}} \frac{\Gamma(2l + 2)}{\Gamma(l + 1 + i\eta)} \left\{ \frac{\Gamma(l + 1 + i\eta)}{\Gamma(l + 1 - i\eta)} \right.$$

$$\left. \times e^{i(kr - \eta \ln(2kr) - \pi(l+1)/2)} + e^{-i(kr - \eta \ln(2kr) - \pi(l+1)/2)} \right\}.$$

The next step involves the definition of the Coulomb phase shift γ_l by

$$e^{2i\gamma_l} = \frac{\Gamma(l + 1 + i\eta)}{\Gamma(l + 1 - i\eta)}, \tag{1.61}$$

as well as factoring out $e^{i\gamma_l}$ and summarising the result in the form

$$F(\eta, kr) \longrightarrow C_l \, \frac{e^{\pi\eta/2} e^{i\gamma_l}}{2^l} \frac{\Gamma(2l + 2)}{\Gamma(l + 1 + i\eta)} \sin(kr - \pi l/2 + \gamma_l - \eta \ln(2kr)).$$

$$\tag{1.62}$$

Normalisation It is standard to choose the normalisation C_l of the radial part of the partial wave expansion of the Coulomb problem in such a way that the regular Coulomb function

$$F_l(\eta, kr) = C_l \, (kr)^{l+1} e^{-ikr} \, F(l+1+i\eta, 2l+2, 2ikr)$$

goes over into the regular solution $R_l(kr)$ of the free radial equation for $\eta = 0$. This agrees with the statement

$$F_l(0, kr) = R_l(kr) = \frac{i^l (2l+1)}{k} u_l(kr) \tag{1.63}$$

in Sect 1.2. The solution of the Coulomb problem with $\eta = 0$

$$F_l(0, kr) = C_l \, (kr)^{l+1} e^{-ikr} \, F(l+1, 2l+2, \, 2ikr)$$

can be rewritten with the relation

$$F(l+1, 2l+2, 2ikr) = \frac{2^{l+1} \Gamma(l+3/2)}{\Gamma(1/2)(kr)^{l+1}} e^{ikr} u_l(kr)$$

as

$$F_l(0, kr) = C_l \, \frac{2^{l+1} \Gamma(l+3/2)}{\Gamma(1/2)} u_l(kr).$$

Comparison with (1.63) yields the normalisation

$$C_l = \frac{i^l (2l+1) \Gamma(1/2)}{2^{l+1} \Gamma(l+3/2) k}. \tag{1.64}$$

Alternative Form of the Coulomb Phase Shift The exponent in (1.61) can be written as

$$2i\gamma_l = \ln \Gamma(l+1+i\eta) - \ln \Gamma(l+1-i\eta).$$

With the use of the relations

$$\Gamma(z^*) = \Gamma(z)^*,$$

$$\ln z = \ln |z| + i \arg z$$

one finds for the phase γ_l

$$\gamma_l = \arg \Gamma(l+1+i\eta). \tag{1.65}$$

1.5.8 The Coulomb Problem in Parabolic Coordinates: Asymptotic Limit

The Kummer function F in $(x = k(r - z))$

$$f_1(r, z) = C\, e^{ikz} F(-i\eta, 1, ix)$$

goes for $kr \to \infty$ over into

$$F(-i\eta, 1, ix) \longrightarrow \frac{\Gamma(1)}{\Gamma(1 + i\eta)} e^{i\pi(-i\eta)} (ix)^{i\eta} + \frac{\Gamma(1)}{\Gamma(-i\eta)} e^{ix} (ix)^{-i\eta-1}.$$

Direct transformation of

$$(ix)^{i\eta} = (e^{i\pi/2})^{i\eta} (e^{\ln x})^{i\eta} = e^{-\eta\pi/2} e^{i\eta \ln kr(1-\cos\theta)}$$

and the formula (see Abramowitz/Stegun, p. 256)

$$\Gamma(-i\eta) = -\frac{\Gamma(1 - i\eta)}{i\eta},$$

as well as

$$(ix)^{-i\eta-1} = \frac{1}{r} \frac{1}{ikr(1 - \cos\theta)} e^{\eta\pi/2} e^{-i\eta \ln k(r-z)},$$

$$e^{-i\eta \ln k(r-z)} = e^{-i\eta \ln kr} e^{-i\eta(\ln k(r-z)-\ln kr)} = e^{-i\eta \ln kr} e^{-i\eta \ln(1-\cos\theta)}$$

yield

$$f_1(r, z) \longrightarrow \frac{C\, e^{\eta\pi/2}}{\Gamma(1 + i\eta)} \left\{ e^{i(kz+\eta \ln k(r-z))} \right.$$

$$\left. - \frac{e^{i(kr-\eta \ln kr)}}{r} \frac{\Gamma(1 + i\eta)}{\Gamma(1 - i\eta)} \frac{\eta e^{-i\eta \ln(1-\cos\theta)}}{k(1 - \cos\theta)} \right\}.$$

If one inserts in this relation the definition of the phase shift γ_0 and uses the normalisation

$$C = e^{-\eta\pi/2} \Gamma(1 + i\eta),$$

one finds

$$f_1(r, z) \longrightarrow \left\{ e^{i(kz + \eta \ln k(r-z))} \right.$$

$$\left. + \frac{e^{i(kr - \eta \ln kr)}}{r} \left[\frac{-\eta \, e^{i(2\gamma_0 - \eta \ln(1 - \cos\theta))}}{k(1 - \cos\theta)} \right] \right\} .$$

Literature Quoted in the Preliminary Remarks and in Chap. 1

1. H. Geiger and E. Marsden, Phil. Mag. **25**, p. 604 (1913)
2. E. Rutherford, Phil. Mag. **21**, p. 669 (1911).
3. R. M. Dreizler, C. S. Lüdde: Theoretical Physics 1, Theoretical Mechanics. Springer Verlag, Heidelberg (2002 and 2008)
4. N. Levinson, Danske Videnskab. Selskab, Mat.-fys. Medd. **25**, No 9 (1949)
5. S. Flügge: Practical Quantum Mechanics. SpringerVerlag, Heidelberg (1974)
6. M. Abramowitz, I. Stegun: Handbook of Mathematical Functions. Dover Publications, New York (1974)
7. P. Moon, D. Eberle: Field Theory Handbook. Springer Publishing House, Heidelberg (1961)
8. J. R. Taylor: Scattering Theory, The Quantum Theory of Nonrelativistic Collisions. John Wiley, New York (1972)

Elastic Scattering: Stationary Formulation—Integral Equations

2

An alternative to the discussion of potential scattering problems on the basis of differential equations is the treatment of such problems in terms of corresponding integral equations. An advantage of using integral equations is the fact that boundary conditions can be incorporated explicitly into the equations. This allows a compact formal discussion of potential scattering problems as well as access to more complex collision processes. The simplest form of the integral equations for the potential scattering problem will be presented in the next sections. The formal device used in this discussion is the standard Dirac notation of quantum mechanics.

The transition from a differential equation for the scattering wave function to an integral equation requires the introduction of Green's functions. These functions are actually *distributions* defined in the complex wave number plane by contour integration. Different boundary conditions for the problem of interest can be introduced via a choice of different contours. The integral equations obtained in this fashion are called Lippmann-Schwinger equations.[1] Of particular interest, however, are not the scattering wave functions but the scattering amplitudes. In order to discuss these in a direct fashion, one defines the T-matrix operator or the associated T-matrix elements.[2] Integral equations for the T-matrix elements can be obtained from the integral equations for the wave functions. Since the solution of integral equations is not easily accessible, these equations are often used as a starting point for the formulation of approximations, which can be discussed and evaluated both in analytical form and with graphical methods. The chapter concludes with a discussion of the optical theorem from the point of view of the T-matrix.

---

[1] B. A. Lippmann, Phys. Rev. Lett. **79**, p. 461 (1950).

[2] T stands for transition.

© Springer-Verlag GmbH Germany, part of Springer Nature 2022
R. M. Dreizler et al., *Quantum Collision Theory of Nonrelativistic Particles*,
https://doi.org/10.1007/978-3-662-65591-7_2

2.1   The Lippmann-Schwinger Equation for the Scattering Wave Function

The starting point is the Schrödinger equation in the form

$$(E_0 - \hat{H}_0)|\psi\rangle = \hat{V}|\psi\rangle, \tag{2.1}$$

where $\hat{H}_0$ stands in most cases for the kinetic energy

$$\hat{H}_0 \equiv \hat{T}.$$

The energy E_0 is given as $E_0 = (\hbar k_0)^2/2m_0$. The transition to an equivalent integral equation involves the following steps: First, one interprets the right-hand side of (2.1) as the inhomogeneous term of an inhomogeneous differential equation. The solution of the homogeneous differential equation is a plane wave

$$|\psi_{\text{hom}}\rangle = |k_0\rangle$$

with the representation in coordinate space (and standard normalisation)

$$\langle r|k_0\rangle = \psi_{k_0}(r) = \frac{1}{(2\pi)^{3/2}}e^{ik_0\cdot r}.$$

For a representation of a particular solution of the inhomogeneous differential equation one uses the resolvent or Green's function $\hat{G}_0(E_0)$ given by

$$(E_0 - \hat{H}_0)\,\hat{G}_0(E_0) = 1 \quad \text{or} \quad \hat{G}_0(E_0) = \frac{1}{(E_0 - \hat{H}_0)}. \tag{2.2}$$

A particular integral of (2.1) is

$$|\psi_{\text{part}}\rangle = \hat{G}_0(E_0)\hat{V}|\psi\rangle.$$

The formal solution of (2.1)

$$|\psi\rangle = |k_0\rangle + \hat{G}_0(E_0)\hat{V}|\psi\rangle \tag{2.3}$$

corresponds to an integral equation, the *Lippmann-Schwinger equation* for the state $|\psi\rangle$. In the coordinate space representation, it is an integral equation for the wave function

$$\langle r|\psi\rangle = \langle r|k_0\rangle + \int d^3r' \int d^3r'' \langle r|\hat{G}_0(E_0)|r'\rangle\langle r'|\hat{V}|r''\rangle\langle r''|\psi\rangle, \tag{2.4}$$

which for a *local* potential with

$$\langle r'|\hat{V}|r''\rangle = \delta(r' - r'')v(r')$$

simplifies to

$$\psi(r) = \psi_{k_0}(r) + \int d^3r' G_0(r, r'; E_0)v(r')\psi(r'). \tag{2.5}$$

2.1.1 Green's Functions

The *kernel* of the integral equation (2.4), the Green's function, is, however, not completely defined by (2.2). One can show this directly via the explicit representation in coordinate space. It is (set $\hat{H}_0 = \hat{T}$)

$$
\begin{aligned}
G_0(r, r'; E_0) \equiv G_0(r, r') &= \left\langle r\left|\frac{1}{(E_0 - \hat{T})}\right|r'\right\rangle \\
&= \int d^3k \int d^3k' \langle r|k\rangle \left\langle k\left|\frac{1}{(E_0 - \hat{T})}\right|k'\right\rangle \langle k'|r'\rangle \\
&= \frac{2m_0}{(2\pi)^3\hbar^2} \int d^3k \frac{e^{ik\cdot(r-r')}}{(k_0^2 - k^2)} \\
&= \frac{m_0}{2\pi^2\hbar^2 |r - r'|} \int_{-\infty}^{+\infty} kdk \frac{\sin k |r - r'|}{(k_0^2 - k^2)}.
\end{aligned}
\tag{2.6}
$$

The integrand indicates that the Green's function in (2.4) is singular at the points $k = \pm k_0$. However, it is possible to extend the definition and obtain a regular Green's function. It is also possible to demonstrate that the solution of the integral equation with the extended Green's function satisfies the required boundary condition (1.6)

$$\psi_{\text{asymp}}(r) \xrightarrow{r \to \infty} N\left\{e^{ik_0\cdot r} + f(\theta, \varphi)\frac{e^{ik_0 r}}{r}\right\}.$$

The extension consists in a suitable shift of the position of the poles from the real axis into the complex k-plane and the replacement of the integration along the real axis by a Cauchy contour integration over a closed curve in that plane. For this one uses the definitions[3]

$$\hat{G}_0^{(\pm)}(E_0) \equiv \hat{G}_0^{(\pm)} = \lim_{\epsilon \to 0} \frac{1}{(E_0 - \hat{T} \pm i\epsilon)} \tag{2.7}$$

---

[3] The dependence on the energy E_0 is not always displayed.

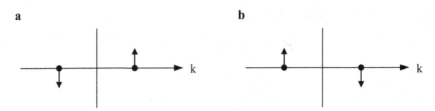

Fig. 2.1 Pole shifts for (**a**) $\hat{G}_0^{(+)}$ and (**b**) $\hat{G}_0^{(-)}$

and a suitable choice of the integration contour. The corresponding Green's functions[4] are the asymptotically outgoing (+) and the incoming (−) spherical Green's function. They describe the free motion including the specified boundary condition.

The extension of the Green's function in (2.4) requires the shifts of the position of the poles, as indicated in Figs. 2.1a, b, as well as a change of the integration along the real k-axis to a suitable contour in the entire complex k-plane. The final result is obtained by taking the limit $\epsilon \to 0$.

In order to check that this prescription yields the correct result, one computes the coordinate space representation of the Green's functions $G_0^{(\pm)}(r, r')$. The steps of the calculation follow the procedure indicated in Eq. (2.6) and are detailed in Sect. 2.4.1. The ansatz[5] is

$$
\begin{aligned}
G_0^{(\pm)}(r, r') &= \lim_{\epsilon \to 0} \left\langle r \left| \frac{1}{(E_0 - \hat{T} \pm i\epsilon)} \right| r' \right\rangle \\
&= \lim_{\epsilon \to 0} \left[\frac{m_0}{4i\pi^2\hbar^2 |r - r'|} \int_{-\infty}^{+\infty} k\, dk\, \frac{(e^{ik\,|r-r'|} - e^{-ik\,|r-r'|})}{(k_0^2 - k^2 \pm i\epsilon)} \right].
\end{aligned}
$$
(2.8)

The argument of the exponential function in the first integral in (2.8) allows a complementation of the integral along the real axis to a contour integral with an (infinitely large) semicircle in the upper half-plane. In the second integral, the complement is a corresponding semicircle in the lower half-plane. The complements extend the integrals along the real axis to integrals over closed contours *without* changing the value of the integrals. In the first integral in (2.8), the right hand pole is located in the contour for the Green's function $G_0^{(+)}$, for the second integral the pole on the left side has been shifted into the contour (Figs. 2.2a, b). With the Cauchy formula

$$
\oint dz\, \frac{f(z)}{(z - z_0)} = 2\pi i f(z_0)
$$

---

[4] The official names are: retarded (+) and advanced Green's function (−).

[5] In the second line $\epsilon' = 2m_0\epsilon/\hbar^2$ is changed to ϵ as only the limiting value $\epsilon \to 0$ plays a role.

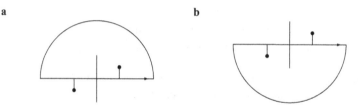

Fig. 2.2 Contributions to the contour integration for $G_0^{(+)}(r, r')$

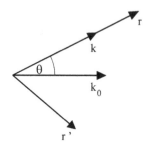

Fig. 2.3 Vectors for the wave numbers k and k_0

follows—after identification of the position of the poles, the direction of the contour integration and the limiting process—for the retarded Green's function

$$G_0^{(+)}(r, r') = -\frac{m_0}{2\pi\hbar^2}\frac{e^{ik_0|r-r'|}}{|r-r'|}. \tag{2.9}$$

The final step of the verification is the extraction of the asymptotic limit. One expands in (2.9)

$$|r - r'| = [r^2 - 2rr'\cos\theta_{r,r'} + r'^2]^{1/2} \quad \text{and} \quad \frac{1}{|r-r'|}$$

to leading order

$$|r - r'| \xrightarrow{r \gg r'} (r - r'\cos\theta_{r,r'}) + O(1/r),$$

$$\frac{1}{|r-r'|} \xrightarrow{r \gg r'} \frac{1}{r} + O(1/r^2)$$

and defines a wave number vector k with the properties (Fig. 2.3)

- $|k|^2 = |k_0|^2$ $\rightarrow$ the vector has the same length as the vector k_0,
- $k \cdot r = k_0 \, r$ $\rightarrow$ the vector has the same direction as the position vector r,
- $k \cdot k_0 = k_0^2 \cos \theta \rightarrow$ the vector k marks the scattering angle θ.

With these definitions one finds for the asymptotic form of the Green's function with the boundary condition of outgoing spherical waves the asymptotic limit

$$G_0^{(+)}(r, r') \xrightarrow{r \to \infty} - \frac{m_0}{2\pi \hbar^2} \frac{1}{r} e^{ik_0 r} e^{-ik \cdot r'} \tag{2.10}$$

and thus for the corresponding wave function (2.5)

$$\psi_{k_0}^{(+)}(r) \xrightarrow{r \to \infty} \frac{1}{(2\pi)^{3/2}} \left\{ e^{ik_0 \cdot r} + \right.$$
$$\left. \left[-\frac{m_0}{\hbar^2} (2\pi)^{1/2} \int d^3 r' \, e^{-ik \cdot r'} \, v(r') \psi_{k_0}^{(+)}(r') \right] \frac{e^{ik_0 r}}{r} \right\}. \tag{2.11}$$

The wave function has the required asymptotic form, if one identifies the expression in the square brackets with the scattering amplitude

$$f(\theta, \varphi) = \left[-\frac{m_0}{\hbar^2} (2\pi)^{1/2} \int d^3 r' \, e^{-ik \cdot r'} v(r') \psi_{k_0}^{(+)}(r') \right]. \tag{2.12}$$

This result also states: In order to calculate the scattering amplitude, one has to know the wave function in the complete space (respectively, in the part of the space where the potential differs from the value zero).

In summary of these arguments one can note the following statements:

- The integral equation for the scattering wave function $\psi_{k_0}^{(+)}(r)$, which satisfies the required boundary condition, has the form (using Dirac notation)

$$|\psi_{k_0}^{(+)}\rangle = |k_0\rangle + \hat{G}_0^{(+)} \hat{V} |\psi_{k_0}^{(+)}\rangle. \tag{2.13}$$

Given are the direction of incidence k_0 with the energy $E_0 = (\hbar k_0)^2/(2m_0)$ and the boundary conditions, which are indicated by the index $(+)$.
- The operator for the Green's functions with free propagation of a particle in the form of outgoing *spherical waves* is[6]

$$\hat{G}_0^{(+)} = \frac{1}{(E_0 - \hat{T} + i\epsilon)}. \tag{2.14}$$

---

[6] The limit $\epsilon \to 0$ is not always indicated explicitly, but is implied.

- The scattering amplitude for a local potential can be calculated by

$$f(\theta, \varphi) = -\frac{m_0}{\hbar^2} (2\pi)^2 \langle \mathbf{k} | \hat{V} | \psi_{\mathbf{k}_0}^{(+)} \rangle. \tag{2.15}$$

The vectors of equal length $\mathbf{k}_0$ and $\mathbf{k}$ mark the scattering angles θ and φ (respectively only θ in the case of cylindrical symmetry). Equation (2.15) is also valid for nonlocal potentials.

The Green's function with the boundary condition of an incoming spherical wave

$$G_0^{(-)}(\mathbf{r}, \mathbf{r}') = \langle \mathbf{r} | \hat{G}_0^{(-)} | \mathbf{r}' \rangle = -\frac{m_0}{2\pi\hbar^2} \frac{e^{-ik_0|\mathbf{r}-\mathbf{r}'|}}{|\mathbf{r}-\mathbf{r}'|} \tag{2.16}$$

is obtained with the pole shifts indicated in Fig. 2.1b and the formal definition

$$\hat{G}_0^{(-)} = \frac{1}{(E_0 - \hat{T} - i\epsilon)} \tag{2.17}$$

by means of an analogous argumentation. The associated wave function is determined by the integral equation

$$|\psi_{\mathbf{k}_0}^{(-)}\rangle = |\mathbf{k}_0\rangle + \hat{G}_0^{(-)} \hat{V} |\psi_{\mathbf{k}_0}^{(-)}\rangle.$$

The functions $G_0^{(-)}(\mathbf{r}, \mathbf{r}')$ and $G_0^{(+)}(\mathbf{r}, \mathbf{r}')$ can be interpreted as time-reversed partners.

From the Green's functions for incoming and outgoing waves one obtains with the definition

$$\hat{G}_0^{(s)}(E_0) = \frac{1}{2} \left\{ \hat{G}_0^{(+)}(E_0) + \hat{G}_0^{(-)}(E_0) \right\} \tag{2.18}$$

a Green's function with the boundary condition of a standing wave. The representation in coordinate space is

$$G_0^{(s)}(\mathbf{r}, \mathbf{r}') = -\frac{m_0}{2\pi\hbar^2} \frac{\cos k_0|\mathbf{r}-\mathbf{r}'|}{|\mathbf{r}-\mathbf{r}'|}.$$

In addition to these three Green's functions for the free propagation of particles, one can introduce the *exact Green's function* of a more general Hamiltonian operator $\hat{H} = \hat{H}_0 + \hat{V}$, which describes the propagation of particles under the influence of a potential $\hat{V}$. The operator $\hat{H}_0$ stands here for the kinetic energy or represents a more general one-particle operator. The exact Green's function is obtained by evaluating the operator identity

$$\frac{1}{\hat{A}} = \frac{1}{\hat{B}} + \frac{1}{\hat{B}}(\hat{B} - \hat{A})\frac{1}{\hat{A}} = \frac{1}{\hat{B}} + \frac{1}{\hat{A}}(\hat{B} - \hat{A})\frac{1}{\hat{B}} \qquad (2.19)$$

with

$$\hat{A} = (E_0 - \hat{T} - \hat{V} \pm i\epsilon) \quad \text{and} \quad \hat{B} = (E_0 - \hat{T} \pm i\epsilon)$$

respectively for

$$\hat{A} = (E_0 - \hat{H}_0 - \hat{V} \pm i\epsilon) \quad \text{and} \quad \hat{B} = (E_0 - \hat{H}_0 \pm i\epsilon).$$

The relation

$$\frac{1}{(E_0 - \hat{H}_0 - \hat{V} \pm i\epsilon)} = \frac{1}{(E_0 - \hat{H}_0 \pm i\epsilon)} \\ + \frac{1}{(E_0 - \hat{H}_0 - \hat{V} \pm i\epsilon)}\hat{V}\frac{1}{(E_0 - \hat{H}_0 \pm i\epsilon)} \qquad (2.20)$$

is the formal version of an integral equation for the exact Green's functions

$$\hat{G}^{(\pm)}(E_0) \equiv \hat{G}^{(\pm)} = \frac{1}{(E_0 - \hat{H}_0 - \hat{V} \pm i\epsilon)} = \frac{1}{(E_0 - \hat{H} \pm i\epsilon)}$$

of the Hamiltonian $\hat{H} = \hat{H}_0 + \hat{V}$. It describes the motion in a given potential and the specified boundary conditions. The Eq. (2.20)

$$\hat{G}^{(\pm)}(E_0) = \hat{G}_0^{(\pm)}(E_0) + \hat{G}^{(\pm)}(E_0)\hat{V}\hat{G}_0^{(\pm)}(E_0)$$

or alternatively

$$\hat{G}^{(\pm)}(E_0) = \hat{G}_0^{(\pm)}(E_0) + \hat{G}_0^{(\pm)}(E_0)\hat{V}\hat{G}^{(\pm)}(E_0) \qquad (2.21)$$

are the *Dyson equations* for the exact Green's functions. They allow in principle the computation of $\hat{G}^{(\pm)}$ for a given potential. The fact that knowledge of the exact Green's function corresponds to a solution of the Lippmann-Schwinger equation for the wave function of the scattering problem is shown by the following argument. One replaces in Eq. (2.13) the Green's function $\hat{G}_0^{(+)}$ with the help of the Dyson equation (2.21). In the intermediate result

$$|\psi_{k_0}^{(+)}\rangle = |k_0\rangle + \hat{G}^{(+)}(E_0)\hat{V}\left\{1 - \hat{G}_0^{(+)}(E_0)\hat{V}\right\}|\psi_{k_0}^{(+)}\rangle$$

one recognises on the right side the plane wave state—in the expression in curly brackets applied to the scattering state—so that one can sort the intermediate result in the form

$$|\psi_{k_0}^{(+)}\rangle = \left\{1 + \hat{G}^{(+)}(E_0)\hat{V}\right\}|k_0\rangle. \tag{2.22}$$

In the same fashion one can obtain the integral equation

$$|\psi_{k_0}^{(-)}\rangle = \left\{1 + \hat{G}^{(-)}(E_0)\hat{V}\right\}|k_0\rangle. \tag{2.23}$$

These formal solutions of the exact Lippmann-Schwinger equation state that knowledge of $\hat{G}^{(\pm)}$ allows the calculation of the associated scattering wave functions by simple integration.

A useful form of Eqs. (2.22) and (2.23) can be obtained by rewriting the first term in the curly brackets with the steps

$$\begin{aligned}|\psi_{k_0}^{(\pm)}\rangle &= \lim_{\epsilon \to 0} \hat{G}^{(\pm)}(E_0)\left\{E_0 - \hat{H} \pm i\epsilon + \hat{V}\right\}|k_0\rangle \\ &= \lim_{\epsilon \to 0}\left\{\frac{\pm i\epsilon}{E_0 - \hat{H} \pm i\epsilon}\right\}|k_0\rangle.\end{aligned} \tag{2.24}$$

An alternative to the extension of the definition of Green's functions by shifting the poles, described above, is the inclusion or exclusion of the singular points in the complex k-plane by small semicircles (Fig. 2.4). The result is the same as before, due to the properties of complex analytic functions. It is also possible to compute the spatial representation of Green's functions directly, i.e. without applying the Cauchy formula (details in Sect. 2.4.2) by explicitly specifying the integration path. In this way one obtains the much used principal value formula

$$\begin{aligned}\hat{G}_0^{(\pm)}(E_0) &= \lim_{\epsilon \to 0} \frac{1}{(E_0 - \hat{H}_0 \pm i\epsilon)} \\ &= \mathscr{P}\left\{\frac{1}{(E_0 - \hat{H}_0)}\right\} \mp i\pi\delta(E_0 - \hat{H}_0).\end{aligned} \tag{2.25}$$

The formula (2.25), which is known as *Dirac Identity*, explicitly demonstrates that Green's functions are not functions but distributions. The formal expressions or the

Fig. 2.4 Variation of the contour integration

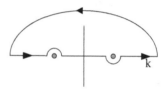

corresponding space or momentum representations only make sense if they occur within an integral, so for example in the integral

$$\int d^3k \ G_0^{(\pm)}(k) f(k),$$

where a regular integrand $f(k)$ and the momentum representation

$$\langle k|\hat{G}_0^{(\pm)}|k'\rangle = \delta(k - k')G_0^{(\pm)}(k)$$

should be used. On the basis of the definitions, one can demonstrate again the property

$$\hat{G}_0^{(\pm)}(E_0)^\dagger = \hat{G}_0^{(\mp)}(E_0),$$

which is evident from Eq. (2.7). A prerequisite for this relation is $\hat{H}_0^\dagger = \hat{H}_0$.

2.2   The Lippmann-Schwinger Equation for the T-Matrix

The concepts and definitions introduced in the previous section allow an extension of the collision theory. Starting from Eq. (2.15) for the scattering amplitude, one defines an operator $\hat{\mathsf{T}}$ with the property[7]

$$\hat{\mathsf{T}}|k_0\rangle = \hat{V}|\psi_{k_0}^{(+)}\rangle. \tag{2.26}$$

The relevant information concerning the scattering is shifted from the scattering state to an operator. The scattering amplitude (2.15) is then written in the form

$$f(\theta, \varphi) = -\frac{m_0}{\hbar^2}(2\pi)^2 \langle k|\hat{\mathsf{T}}|k_0\rangle|_{k=k_0}. \tag{2.27}$$

It is proportional to the matrix element of the operator $\hat{\mathsf{T}}$, the *T-matrix element*, with respect to the plane wave states $|k_0\rangle$ and $|k\rangle$. The vector k_0 marks the direction of the incident beam and the vector k the direction from the target to the detector. The scattering amplitude is therefore a probability measure for the transition from the state $|k_0\rangle$ to the state $|k\rangle$. This justifies the name of the operator $\hat{\mathsf{T}}$ as a transition operator. For elastic scattering of two particles, the energy is conserved ($k = k_0$). In this case one refers to the T-matrix element as an *on-shell T-matrix element* or a *T-matrix element on the energy shell*.

---

[7] This operator should not be confused with the operator for the kinetic energy. For this reason, it is set in *sans serif*.

As the T-matrix elements are more directly related to experiment than the scattering wave function, it is useful to ask whether one can obtain these quantities without first calculating the wave function. In order to find the appropriate equation, one multiplies the Lippmann-Schwinger equation (2.13) by $\hat{V}$ and uses the definition (2.26) of the T-operator

$$\hat{T}|k_0\rangle = \hat{V}|k_0\rangle + \hat{V}\hat{G}_0^{(+)}(E_0)\hat{T}|k_0\rangle.$$

This relation corresponds to the operator equation

$$\hat{T} = \hat{V} + \hat{V}\hat{G}_0^{(+)}(E_0)\hat{T} = \hat{V} + \hat{V}\frac{1}{(E_0 - \hat{H}_0 + i\epsilon)}\hat{T}. \tag{2.28}$$

This integral equation is the Lippmann-Schwinger equation for the T operator. In the momentum representation this is an explicit integral equation for the T-matrix elements

$$\langle k_1|\hat{T}|k_2\rangle = \langle k_1|\hat{V}|k_2\rangle + \frac{2m_0}{\hbar^2}$$
$$\times \lim_{\epsilon \to 0} \int d^3k' \langle k_1|\hat{V}|k'\rangle \frac{1}{(k_0^2 - k'^2 + i\epsilon)} \langle k'|\hat{T}|k_2\rangle. \tag{2.29}$$

One can obtain some variants of (2.28), e.g. by inserting the relation (2.22) for the scattering state into the definition (2.26) and by extracting the operator equation

$$\hat{T} = \hat{V} + \hat{V}\hat{G}^{(+)}(E_0)\hat{V}. \tag{2.30}$$

If one multiplies (2.30) from the left with $\hat{G}_0^{(+)}$, one finds with (2.20)

$$\hat{G}_0^{(+)}\hat{T} = (\hat{G}_0^{(+)} + \hat{G}_0^{(+)}\hat{V}\hat{G}^{(+)})\hat{V} = \hat{G}^{(+)}\hat{V}. \tag{2.31}$$

In the same fashion one obtains by multiplying (2.30) from the right with $\hat{G}_0^{(+)}$

$$\hat{T}\hat{G}_0^{(+)} = \hat{V}(\hat{G}_0^{(+)} + \hat{G}_0^{(+)}\hat{V}\hat{G}^{(+)}) = \hat{V}\hat{G}^{(+)}. \tag{2.32}$$

If the product $\hat{V}\hat{G}^{(+)}$ in (2.21) is replaced by $\hat{T}\hat{G}_0^{(+)}$ this relation leads to

$$\hat{G}^{(+)} = \hat{G}_0^{(+)} + \hat{G}_0^{(+)}\hat{T}\hat{G}_0^{(+)}. \tag{2.33}$$

Each of these relations indicates that knowledge of $\hat{G}^{(+)}$ and of $\hat{T}$ is completely equivalent.

The on-shell and off-shell structure of T-matrix elements is more subtle than these remarks imply. In a matrix element

$$\langle k_1|\hat{T}(E_2)|k_3\rangle$$

occur in fact three energy values, where the energy E_2 is introduced by the Green's function in (2.28). One can therefore distinguish the following cases:

- If all three energy values are equal

$$E_1 = \frac{\hbar^2 k_1^2}{2m_0} = E_2 = E_3,$$

  one speaks of a T-matrix element on the energy shell. The *on-shell* matrix elements are appropriate for the description of elastic scattering.
- Matrix elements with either

$$E_1 = E_2 \neq E_3, \qquad E_2 = E_3 \neq E_1, \qquad E_1 = E_3 \neq E_2$$

  are called *half-off-shell* matrix elements. The Lippmann-Schwinger equation (2.28) shows that these matrix elements are needed for the calculation of on-shell matrix elements.
- If all energy values are different

$$E_1 \neq E_2, \qquad E_1 \neq E_3, \qquad E_2 \neq E_3,$$

  the matrix elements are not on the energy shell, they are *off-shell*.

The energy of the *transition* is not conserved for matrix elements with $E_1 \neq E_3$. In addition to their role in the solution of the Lippmann-Schwinger equation, off-shell matrix elements play a role in some particular physical problems. An example is the *inverse scattering problem,* in which one attempts to reconstruct a potential from the data of scattering experiments (e.g. a nuclear potential from nucleon-nucleon scattering). The proton-proton system (like the neutron-neutron system) has no bound states. In the evaluation of a potential matrix element

$$\text{nonlocal} \quad \langle r|\hat{V}|r'\rangle \qquad \text{or} \quad \text{local} \quad \langle r|\hat{V}|r'\rangle = \delta(r - r')\,v(r)$$

with the completeness relations

$$\int d^3k\ |\psi_k^{(+)}\rangle\langle\psi_k^{(+)}| = \hat{1} \quad \text{and} \quad \int d^3k\ |k\rangle\langle k| = \hat{1}$$

one recognises that both on-shell and off-shell T-matrix elements occur, as one has

$$\langle r|\hat{V}|r'\rangle = \int\int d^3k\, d^3k'\, \langle r|k\rangle\langle k|\hat{V}|\psi_{k'}^{(+)}\rangle\langle\psi_{k'}^{(+)}|r'\rangle$$
$$= \int\int d^3k\, d^3k'\, \langle r|k\rangle\langle k|\hat{T}|k'\rangle\langle\psi_{k'}^{(+)}|r'\rangle.$$

If one uses only elastic scattering data, one cannot distinguish interactions with different off-shell behaviour.

2.2.1   Methods for the Approximate Calculation of T-Matrix Elements

The exact solution of the Lippmann-Schwinger equation for the T-matrix elements is not a simple task. For this reason, various methods for their approximate determination have been developed. A selection of these methods is presented here.

Iteration of the Lippmann-Schwinger equation yields (in the momentum representation)

$$\langle k_1|\hat{T}|k_2\rangle = \langle k_1|\hat{V}|k_2\rangle + \langle k_1|\hat{V}\hat{G}_0^{(+)}\hat{V}|k_2\rangle$$
$$+ \langle k_1|\hat{V}\hat{G}_0^{(+)}\hat{V}\hat{G}_0^{(+)}\hat{V}|k_2\rangle + \cdots . \qquad (2.34)$$

This expansion in powers of $\hat{V}$ is called the *Born series*. The first term of the series, the *Born approximation*

$$\langle k_1|\hat{T}|k_2\rangle \approx \langle k_1|\hat{V}|k_2\rangle$$

is often used for estimates. In the evaluation of the additional terms one uses the completeness relation, e.g. for the second term on the right-hand side of Eq. (2.34)

$$\langle k_1|\hat{V}\hat{G}_0^{(+)}\hat{V}|k_2\rangle = \int d^3k_1'\int d^3k_2'\, \langle k_1|\hat{V}|k_1'\rangle\langle k_1'|\hat{G}_0^{(+)}|k_2'\rangle\langle k_2'|\hat{V}|k_2\rangle.$$

This expression simplifies somewhat, as one finds with the momentum representation of the Green's function

$$\langle k_1'|\hat{G}_0^{(+)}|k_2'\rangle = \lim_{\epsilon\to 0}\left\langle k_1'\left|\frac{1}{(E_0-\hat{T}+i\epsilon)}\right|k_2'\right\rangle$$
$$= \frac{2m_0}{\hbar^2}\lim_{\epsilon'\to 0}\frac{1}{(k_0^2-(k_1')^2+i\epsilon')}\delta(k_1'-k_2').$$

The double integration in the second term of (2.34) is reduced to one integration.

The Born series is the starting point for a variety of approximations in scattering theory. A very useful tool is the two-potential formula, which is the basis of an approximation with *distorted waves*—(Distorted Wave Born Approximation, short *DWBA*). The starting point for the discussion is a Hamiltonian, which can be written in the form

$$\hat{H} = \hat{T} + \hat{V}_1 + \hat{V}_2.$$

The splitting of the potential into two parts can either arise due to the character of the problem or can be introduced arbitrarily. The expectation is that the contribution of the potential $\hat{V}_1$ can be treated exactly, while the first Born approximation or the first terms of a Born series in $\hat{V}_2$ are sufficient for the treatment of the second term.

The explicit derivation begins with the definition (2.26)

$$\langle k'|\hat{T}|k\rangle = \langle k'|\hat{V}|\psi_k^{(+)}\rangle.$$

On the right-hand side the state $\langle k'|$ is now replaced by the Lippmann-Schwinger equation for the potential $\hat{V}_1$

$$|\phi_{k'}^{(\pm)}\rangle = |k'\rangle + \hat{G}_0^{(\pm)}\hat{V}_1|\phi_{k'}^{(\pm)}\rangle. \tag{2.35}$$

If the adjoint equation, suitably resolved, is inserted here, the result is

$$\langle k'|\hat{T}|k\rangle = \langle \phi_{k'}^{(-)}|\hat{V}|\psi_k^{(+)}\rangle - \langle \phi_{k'}^{(-)}|\hat{V}_1\hat{G}_0^{(+)}\hat{V}|\psi_k^{(+)}\rangle.$$

One now uses the Lippmann-Schwinger equation (2.13) for the exact scattering wave function

$$\hat{G}_0^{(+)}\hat{V}|\psi_k^{(+)}\rangle = |\psi_k^{(+)}\rangle - |k\rangle$$

and finds

$$\langle k'|\hat{T}|k\rangle = \langle \phi_{k'}^{(-)}|\hat{V}_2|\psi_k^{(+)}\rangle + \langle \phi_{k'}^{(-)}|\hat{V}_1|k\rangle.$$

On the right-hand side the potential $\hat{V}_2 = \hat{V} - \hat{V}_1$ has been introduced. The second term corresponds to the complex conjugate T-matrix element of the potential $\hat{V}_1$. The result

$$\langle k'|\hat{T}|k\rangle = \langle k'|\hat{T}_1|k\rangle + \langle \phi_{k'}^{(-)}|\hat{V}_2|\psi_k^{(+)}\rangle \tag{2.36}$$

is the exact two-potential formula, also known as *Watson's theorem*. The T-matrix element for the potential $\hat{V} = \hat{V}_1 + \hat{V}_2$ can be represented by the T-matrix element of the potential $\hat{V}_1$ and a matrix element of the potential $\hat{V}_2$ with respect to the exact scattering solution $\psi_k^{(+)}$.

If one considers only first-order contributions in $\hat{V}_2$ in the second term, one obtains the formula of the *DWBA*

$$\langle k'|\hat{T}|k\rangle \approx \langle k'|\hat{T}_1|k\rangle + \langle\phi_{k'}^{(-)}|\hat{V}_2|\phi_k^{(+)}\rangle. \tag{2.37}$$

In the second term only solutions of the $\hat{V}_1$ problem (the distorted waves) occur and just one interaction term due to the potential $\hat{V}_2$.

In order to see the detailed structure of the second term in (2.37), one inserts the Lippmann-Schwinger equation for the scattering states in the potential $\hat{V}_1$ and obtains

$$\langle\phi_{k'}^{(-)}|\hat{V}_2|\phi_k^{(+)}\rangle = \langle k'|\hat{V}_2|k\rangle + \int d^3k'' \langle\phi_{k'}^{(-)}|\hat{V}_1\hat{G}_0^{(+)}|k''\rangle\langle k''|\hat{V}_2|k\rangle$$

$$+ \int d^3k'' \langle k'|\hat{V}_2|k''\rangle\langle k''|\hat{G}_0^{(+)}\hat{V}_1|\phi_k^{(+)}\rangle$$

$$+ \int\int d^3k''d^3k''' \langle\phi_{k'}^{(-)}|\hat{V}_1\hat{G}_0^{(+)}|k''\rangle\langle k''|\hat{V}_2|k'''\rangle$$

$$\times \langle k'''|\hat{G}_0^{(+)}\hat{V}_1|\phi_k^{(+)}\rangle.$$

For the Green's function $\hat{G}_0^{(+)}$ one has

$$\langle k_1|\hat{G}_0^{(+)}|k_2\rangle = \delta(k_1 - k_2)G_0^{(+)}(k_1),$$

so that, using the definition of the T-matrix of the potential $\hat{V}_1$, one obtains the result

$$\langle\phi_{k'}^{(-)}|\hat{V}_2|\phi_k^{(+)}\rangle = \langle k'|\hat{V}_2|k\rangle + \int d^3k'' \langle k'|\hat{T}_1|k''\rangle G_0^{(+)}(k'')\langle k''|\hat{V}_2|k\rangle$$

$$+ \int d^3k'' \langle k'|\hat{V}_2|k''\rangle G_0^{(+)}(k'')\langle k''|\hat{T}_1|k\rangle$$

$$+ \int\int d^3k''d^3k''' \langle k'|\hat{T}_1|k''\rangle G_0^{(+)}(k'')\langle k''|\hat{V}_2|k'''\rangle \tag{2.38}$$

$$\times G_0^{(+)}(k''')\langle k'''|\hat{T}_1|k\rangle.$$

A contribution of first order in the potential $\hat{V}_2$ is coupled in all possible ways to the T-matrix elements of the potential $\hat{V}_1$.

The DWBA approximation is not only used to discuss elastic scattering processes by a sum of two potentials, such as the elastic scattering of protons from nuclei by a short-range nuclear potential and a long-range Coulomb potential. The DWBA is also a much used tool in the analysis of inelastic processes such as nucleon transfer processes. As a concrete example, the (d–p) stripping reaction, in which a neutron is transferred from a deuteron projectile to nuclei, is addressed in Sect. 7.6.

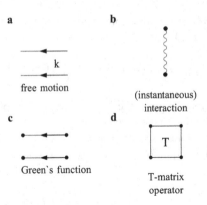

Fig. 2.5 Basic elements of the Feynman diagrams (**a–d**) for the scattering of two particles

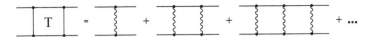

Fig. 2.6 The Born series

The partial wave expansion of Sect. 1.2 is also a useful tool for the calculation and discussion of the T-matrix elements. This topic is treated in Sect. 3.5 after the introduction of the S-matrix and the K-matrix.

2.2.2 Approximation Methods in Terms of Feynman Diagrams

The discussion of the T-matrix, including the discussion of approximations, can alternatively be done in a more pictorial form. One uses for this purpose a graphical representation of the individual elements of the theory in the form of Feynman diagrams[8] with the option to define the graphs either as the scattering of a particle by an external potential (*r* is then the particle coordinate) or as the scattering of *two* particles by an interaction potential (*r* is then the coordinate of the relative motion).

2.2.2.1 Elastic Scattering of Two Particles

The basic elements of the graphical representation for the *scattering of two particles* are shown in Fig. 2.5, where time is supposed to flow in the horizontal direction.

The diagram for the Born series in Fig. 2.6 (without the indices for the characterisation of the plane waves) is a sum of ladders, in other words a multiple scattering expansion. In the second term and in all higher order terms, one integrates over the wave numbers of the (interior) Green's functions.

---

[8] This descriptive technique was introduced in 1949 by R. P. Feynman, Phys. Rev. **76**, p. 749 (1949) for the discussion of electron-positron processes in a transparent way. This technique is used in many branches of physics.

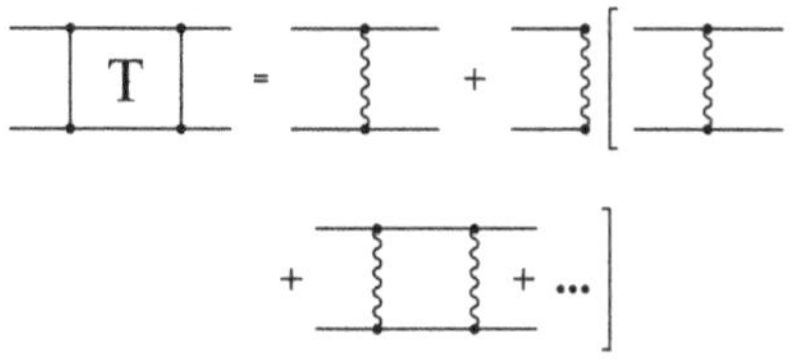

Fig. 2.7 Resummation of the Born series

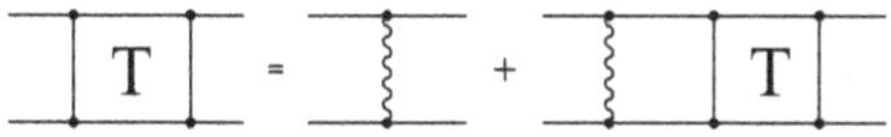

Fig. 2.8 Lippman-Schwinger equation (2.29) for T-matrix elements

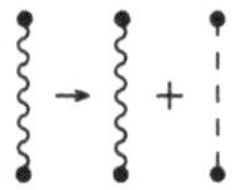

Fig. 2.9 Splitting of the interaction

For a high relative energy, one can expect that the exchange of many *interaction quanta* does not take place. The *first Born approximation*—the contribution of the first term on the right-hand side—

$$\langle k_1|\hat{T}|k_2\rangle_{\text{Born}} = \langle k_1|\hat{V}|k_2\rangle \tag{2.39}$$

is a much used approximation for sufficiently energetic collisions. The reason is the fact that the Fourier transform for local potentials with respect to momentum transfer $\hbar q = \hbar(k_2 - k_1)$

$$\langle k_1|\hat{V}|k_2\rangle = \frac{1}{(2\pi)^3}\int d^3r\, v(r)e^{iq\cdot r} = \bar{v}(q)$$

can often be calculated directly. The evaluation of the higher-order contributions is much more time consuming. The Born series can be resummed in graphical form. As indicated in Figs. 2.7 and 2.8 one interaction on the left-hand side is split off. The sum of the remaining terms on the right-hand side then corresponds to a T-matrix, so that one obtains the Lippmann-Schwinger equation.

In order to discuss the two-potential formula (2.36) in graphical form, one substitutes each interaction $\hat{V}$ as shown in Fig. 2.9 by the sum of the potentials $\hat{V}_1$ and $\hat{V}_2$. If one restricts oneself to the first Born approximation in $\hat{V}_2$, only graphs with *one* $\hat{V}_2$—line have to be considered. They either occur on the incoming or the outgoing side or are embedded within $\hat{V}_1$—lines (Fig. 2.10). Resummation of the contributions of $\hat{V}_1$—lines leads to two terms, each with a T-matrix of the potential $\hat{V}_1$

$$\langle k'|\hat{T}_1|k\rangle$$

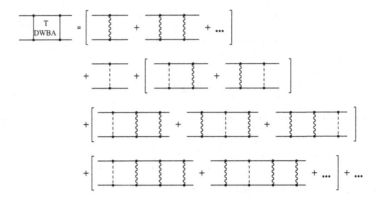

Fig. 2.10 The T-matrix element in the DWBA

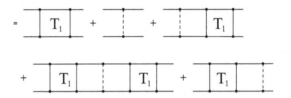

Fig. 2.11 Graphical form of the DWBA

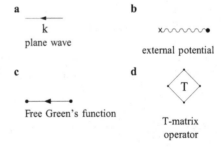

Fig. 2.12 Basic elements (**a**)–(**d**) for the Feynman representation of scattering of one particle by an external potential

and a contribution, in which the $\hat{V}_2$—line is embedded between two T_1—matrices (Fig. 2.11).

2.2.2.2 A Particle Is Scattered by an External Potential

For the graphical representation of Eq. (2.29) one uses the elements shown in Fig. 2.12.

The expansion of the resulting Lippmann-Schwinger equation in terms of powers of the external potential yields a Born series describing the multiple scattering of the particle (Fig. 2.13).

Fig. 2.13 Born series of the single particle scattering problem

2.3   The T-Matrix and the Optical Theorem

The direct relation (2.27) between the scattering amplitude and the T-matrix element leads to the question of the form of the optical theorem in the context of the discussion of the T-matrix. The starting point for an answer to this question is the Lippmann-Schwinger equation in the form

$$\langle \mathbf{k}'|\hat{T}|\mathbf{k}\rangle = \langle \mathbf{k}'|\hat{V}|\mathbf{k}\rangle + \langle \mathbf{k}'|\hat{V}\hat{G}(E_0)\hat{V}|\mathbf{k}\rangle.$$

If one expands the term with the exact Green's function with the completeness relation for the eigenstates of the full Hamiltonian $\hat{H} = \hat{H}_0 + \hat{V}$, bound states must be taken into account (if they occur) in addition to the scattering states

$$\hat{1} = \int \mathrm{d}^3 k |\psi_k^{(+)}\rangle\langle\psi_k^{(+)}| + \sum_n |n\rangle\langle n|.$$

In the result

$$
\begin{aligned}
\langle \mathbf{k}'|\hat{T}|\mathbf{k}\rangle = \langle \mathbf{k}'|\hat{V}|\mathbf{k}\rangle &+ \int \mathrm{d}^3 k'' \frac{\langle \mathbf{k}'|\hat{V}|\psi_{k''}^{(+)}\rangle\langle\psi_{k''}^{(+)}|\hat{V}|\mathbf{k}\rangle}{(E_0 - E(k'') + i\epsilon)} \\
&+ \sum_n \frac{\langle \mathbf{k}'|\hat{V}|n\rangle\langle n|\hat{V}|\mathbf{k}\rangle}{(E_0 - E_n + i\epsilon)}
\end{aligned}
\tag{2.40}
$$

the following simplifications are possible:

- Since the energy of the scattering states E_0 is above 0 and the energy of the bound states E_n is below 0, the addition of $i\epsilon$ in the last term is not necessary.
- The integration in the second term can be replaced, using the definition of the operator $\hat{T}$ in (2.26), by an integration over plane waves

$$\int \mathrm{d}^3 k''\, \hat{V}|\psi_{k''}^{(+)}\rangle\langle\psi_{k''}^{(+)}|\hat{V} = \int \mathrm{d}^3 k''\, \hat{T}|\mathbf{k}''\rangle\langle \mathbf{k}''|\hat{T}^{\dagger}.$$

- As *kernel* of Eq. (2.40) one then obtains a (formal) nonlinear integral equation

$$\hat{T} = \hat{V} + \int \mathrm{d}^3 k'' \frac{\hat{T}|\mathbf{k}''\rangle\langle \mathbf{k}''|\hat{T}^{\dagger}}{(E_0 - E(k'') + i\epsilon)} + \sum_n \frac{\hat{V}|n\rangle\langle n|\hat{V}}{(E_0 - E_n)}
\tag{2.41}
$$

for the T-matrix, which is known as *Low equation*.

The combination $\hat{T} - \hat{T}^\dagger$ yields

$$\hat{T} - \hat{T}^\dagger = \int d^3k'' \, \hat{T}|k''\rangle \left\{ \frac{1}{(E_0 - E(k'')) + i\epsilon} - \frac{1}{(E_0 - E(k'')) - i\epsilon} \right\} \langle k''|\hat{T}^\dagger,$$

as both the potential term ($\hat{V} = \hat{V}^\dagger$) as well as the contribution of the bound states are cancelled. The Dirac identity (2.25) then leads to the final result

$$\hat{T} - \hat{T}^\dagger = -2\pi i \int d^3k'' \, \hat{T}|k''\rangle \, \delta(E_0 - E(k'')) \, \langle k''|\hat{T}^\dagger,$$

or, after rewriting the argument of the δ-function, to

$$i(\hat{T} - \hat{T}^\dagger) = \frac{4\pi m_0}{\hbar^2} \int d^3k'' \, \hat{T}|k''\rangle \, \delta(k_0^2 - (k'')^2) \, \langle k''|\hat{T}^\dagger. \tag{2.42}$$

This relation is a generalisation of the simple optical theorem (1.10). It can be evaluated for on-shell as well as off-shell matrix elements. If one considers the on-shell matrix element with

$$E_0 = E(k) = E(k'),$$

one finds[9]

$$i\langle k|(\hat{T} - \hat{T}^\dagger)|k'\rangle_{\text{on}} = \frac{2\pi m_0 k_0}{\hbar^2} \int d\Omega_{k''} \langle k|\hat{T}|k''\rangle_{\text{on}} \langle k''|\hat{T}^\dagger|k'\rangle_{\text{on}}. \tag{2.43}$$

If one uses the relation (2.27)

$$\langle k|\hat{T}|k'\rangle_{\text{on}} = -\frac{\hbar^2}{4\pi^2 m_0} f(\Omega_{k,k'})$$

with $\theta_k = \theta_{k'} = 0$, $k = k' = k_0$ and $\varphi_k = \varphi_{k'}$ one recovers the simple optical theorem

$$\text{Im} f(0) = \frac{k_0}{4\pi} \int d\Omega_{k''} |f(\Omega_{0,k''})|^2 = \frac{k_0}{4\pi} \sigma.$$

---

[9] For this step of the calculation one uses $\int d^3k'' = \frac{1}{2} \int k'' d(k''^2) d\Omega_{k''}$.

2.4   Detailed Calculations for Chap. 2

2.4.1   Calculation of the Regularised Green's Function

The denominator of the integral

$$I_1 = \int_{-\infty}^{\infty} k \, dk \, \frac{e^{ik|r-r'|}}{(k_0^2 - k^2 + i\epsilon)}$$

can be factorised in the form

$$k_0^2 - k^2 + i\epsilon = -[k - (k_0 + i\epsilon)][k + (k_0 + i\epsilon)].$$

The parameter ϵ was once again renamed $\epsilon \rightarrow 2k_0\epsilon$ and a term in ϵ^2, which vanishes faster in the limiting case, was added. If one changes the integral to a contour integral by means of an infinite semicircle in the upper complex k-plane, the pole shifted upwards at $k = k_0 + i\epsilon$ is included. The contribution of the semicircle has the value zero because of

$$\lim_{|k|\to\infty} e^{ik} = \lim_{|k|\to\infty} \left\{ e^{i|k|\cos\phi} \right\} e^{-|k|\sin\phi} = 0.$$

For the integral I_1 one obtains therefore the Cauchy integral

$$I_1 = -\oint dk \, \frac{k e^{ik|r-r'|}}{(k + k_0 + i\epsilon)} \cdot \frac{1}{(k - (k_0 + i\epsilon))},$$

where the direction of the integration is positive. In the intermediate result

$$I_1 = -2\pi i \frac{(k_0 + i\epsilon)e^{i(k_0+i\epsilon)|r-r'|}}{2(k_0 + i\epsilon)}$$

the limit $\epsilon \rightarrow 0$ can be executed directly. The final result is

$$I_1 = -\pi i e^{ik_0|r-r'|}.$$

For the second integral

$$I_2 = -\int_{-\infty}^{\infty} k dk \, \frac{e^{-ik|r-r'|}}{(k_0^2 - k^2 + i\epsilon)}$$

the same factorisation is used. In this case, the integral must be completed by a semicircle in the lower half plane, so that the downward shifted pole is included at $k = -(k_0 + i\epsilon)$. The Cauchy integral

$$I_2 = \oint dk \, \frac{k e^{-ik|r-r'|}}{(k - (k_0 + i\epsilon))} \cdot \frac{1}{(k + (k_0 + i\epsilon))},$$

where the direction of integration is negative, leads for $\epsilon \to 0$ to the same result as before

$$I_2 = -\pi i e^{ik_0|r-r'|}.$$

Including the prefactors, the Green's function is

$$G_0^{(+)}(r, r') = \frac{m_0}{4i\pi^2\hbar^2|r-r|} \{I_1 + I_2\} = -\frac{m_0}{2\pi\hbar^2} \frac{e^{ik_0|r-r'|}}{|r-r'|}.$$

2.4.2 The Dirac Identity

The starting point is Eq. (2.6) in the form

$$G_0(r, r') = \frac{m_0}{2\pi^2\hbar^2 q} \int_{-\infty}^{+\infty} dk \, \frac{k}{(k_0^2 - k^2)} \sin kq, \tag{2.44}$$

where the quantity q stands for $q = |r - r'|$. In order to calculate $G_0^{(+)}(r, r')$ one uses the integration path shown in Fig. 2.14 with the contributions:

- From $k = -\infty$ to the point $k = -k_0 - \epsilon_1$ along the real axis,
- avoid the singular point $-k_0$ by a semicircle of radius ϵ_1 in the upper half-plane,
- from $k = -k_0 + \epsilon_1$ to the point $k = k_0 - \epsilon_2$ along the real axis,
- bypass the singular point k_0 by a semicircle of radius ϵ_2 in the lower half-plane,
- from $k = k_0 + \epsilon_2$ to the point $k = +\infty$ along the real axis.

The limiting processes ϵ_1, $\epsilon_2 \longrightarrow 0$ are to be performed after the integrations have been carried out.

Fig. 2.14 Path of integration for the calculation of $G_0^{(+)}(r, r')$

To evaluate the components of the integral along the real axis

$$I = \int_{-\infty}^{-k_0-\epsilon_1} dk \ f(k) + \int_{-k_0+\epsilon_1}^{k_0-\epsilon_2} dk \ f(k) + \int_{k_0+\epsilon_2}^{\infty} dk \ f(k)$$

with

$$f(k) = \frac{k}{(k_0^2 - k^2)} \sin kq$$

one uses the decomposition into partial fractions

$$\frac{k}{(k_0^2 - k^2)} = \frac{1}{2} \left\{ \frac{1}{(k_0 - k)} - \frac{1}{(k_0 + k)} \right\}.$$

The decomposition requires a splitting of the integral I into two independent parts $I = I_1 + I_2$ with

$$I_1 = \int_{-\infty}^{-k_0-\epsilon_1} dk \ f_1(k) + \int_{-k_0+\epsilon_1}^{\infty} dk \ f_1(k),$$

$$I_2 = \int_{-\infty}^{k_0-\epsilon_2} dk \ f_2(k) + \int_{k_0+\epsilon_2}^{\infty} dk \ f_2(k).$$

Each of the integrands

$$f_1(k) = -\frac{\sin kq}{2(k_0 + k)} \quad \text{and} \quad f_2(k) = \frac{\sin kq}{2(k_0 - k)}$$

involves one singular point. The substitution $x = k + k_0$ in the first integral leads to

$$I_1 = -\left\{ \int_{-\infty}^{-\epsilon_1} + \int_{+\epsilon_1}^{\infty} \right\} dx \ \frac{\sin q(x - k_0)}{2x}$$

$$= -\left\{ \int_{-\infty}^{-\epsilon_1} + \int_{+\epsilon_1}^{\infty} \right\} dx \ \frac{(\sin qx \cos qk_0 - \cos qx \sin qk_0)}{2x}.$$

The first part contains the principal value integral

$$\left\{ \int_{-\infty}^{-\epsilon_1} + \int_{+\epsilon_1}^{\infty} \right\} dx \ \frac{\sin qx}{2x} = \mathscr{P} \int_{-\infty}^{\infty} dx \ \frac{\sin qx}{2x} = \frac{\pi}{2},$$

the second does not contribute due to the symmetry of the integrand. Therefore one has

$$I_1 = -\frac{\pi}{2} \cos k_0 q.$$

For the integral I_2 one finds with the substitution $x = k_0 - k$

$$I_2 = \mathscr{P} \int_{-\infty}^{\infty} dx \, \frac{\sin q(k_0 - x)}{2x} = \mathscr{P} \int_{-\infty}^{\infty} dx \, \frac{\sin qk_0 \cos qx - \cos qk_0 \sin qx}{2x}$$

and thus the same result

$$I_2 = -\frac{\pi}{2} \cos k_0 q,$$

so that one can write

$$I = \mathscr{P} \int_{-\infty}^{+\infty} dk \, \frac{k}{(k_0^2 - k^2)} \sin kq = -\pi \cos k_0 q.$$

The contributions over the infinitesimal semicircles can be computed with the substitutions

- $k = -k_0 + \epsilon_1 \, e^{i\phi}$ with the lower limit $\phi = \pi$ and the upper limit $\phi = 0$ for the semicircle at $k = -k_0$,
- $k = k_0 + \epsilon_2 \, e^{i\phi}$ with the lower limit $\phi = \pi$ and the upper limit $\phi = 2\pi$ for the semicircle at $k = +k_0$.

The contribution of the first semicircle is then

$$dk = i\epsilon_1 d\phi \, e^{i\phi}, \quad k_0^2 - k^2 \approx 2k_0\epsilon_1 \, e^{i\phi}, \quad \sin qk \approx -\sin qk_0 + q\epsilon_1 \, e^{i\phi} \cos qk_0$$

to first order in ϵ_1

$$\begin{aligned}
I_{\text{left}} &= \oint_{SC_1} dk \, \frac{k \sin qk}{(k_0^2 - k^2)} \\
&= \frac{i}{2} \int_{\pi}^{0} d\phi \left\{ \sin qk_0 - \epsilon_1 \, e^{i\phi} \left(q \cos qk_0 + \frac{\sin qk_0}{k_0} \right) + \cdots \right\} \\
&\longrightarrow -\frac{i\pi}{2} \sin qk_0.
\end{aligned}$$

The contribution of the second semicircle with

$$dk = i\epsilon_2 \, e^{i\phi} d\phi, \quad k_0^2 - k^2 \approx -2k_0\epsilon_2 \, e^{i\phi}, \quad \sin qk \approx \sin qk_0 + q\epsilon_2 \, e^{i\phi} \cos qk_0$$

has the same value

$$
\begin{aligned}
I_{\text{right}} &= \oint_{SC_2} dk \, \frac{k \sin qk}{(k_0^2 - k^2)} \\
&= -\frac{i}{2} \int_{\pi}^{2\pi} d\phi \left\{ \sin qk_0 + \epsilon_2 \, e^{i\phi} \left(q \cos qk_0 + \frac{\sin qk_0}{k_0} \right) + \cdots \right\} \\
&\longrightarrow -\frac{i\pi}{2} \sin qk_0.
\end{aligned}
$$

The overall result for the free Green's function (2.44) with boundary conditions of outgoing waves is in summary

$$
\begin{aligned}
G_0^{(+)}(\boldsymbol{r}, \boldsymbol{r}') &= \frac{m_0}{2\pi \hbar^2 q} \left\{ \mathscr{P} \int_{-\infty}^{+\infty} \frac{dk}{\pi} \, \frac{k}{(k_0^2 - k^2)} \sin kq - i \sin k_0 q \right\} \\
&= -\frac{m_0}{2\pi \hbar^2 q} \left\{ \cos k_0 q + i \sin k_0 q \right\} \\
&= -\frac{m_0}{2\pi \hbar^2 q} e^{ik_0 q}.
\end{aligned} \tag{2.45}
$$

This result shows explicitly the equivalence of the two rules for defining a Green's function with outgoing waves. The verification of the Dirac identity still requires the evaluation of

$$
\begin{aligned}
-i\pi \langle \boldsymbol{r} | \delta(E_0 - \hat{H}_0) | \boldsymbol{r}' \rangle &= -i\pi \int d^3k' \, d^3k \, \langle \boldsymbol{r} | \boldsymbol{k} \rangle \langle \boldsymbol{k} | \delta(E_0 - \hat{H}_0) | \boldsymbol{k}' \rangle \langle \boldsymbol{k}' | \boldsymbol{r}' \rangle \\
&= -\frac{i}{8\pi^2} \int d^3k \, e^{i\boldsymbol{k} \cdot (\boldsymbol{r} - \boldsymbol{r}')} \delta \left(\frac{\hbar^2 k_0^2}{2m_0} - \frac{\hbar^2 k^2}{2m_0} \right) \\
&= -\frac{im_0}{\pi \hbar^2 q} \int_0^\infty k \, dk \, \delta(k_0^2 - k^2) \sin qk \\
&= -\frac{im_0}{\pi \hbar^2 q} \int_{-\infty}^\infty k \, dk \, \delta(k_0^2 - k^2) \sin qk \\
&= -\frac{im_0}{2\pi \hbar^2 q} \sin qk_0.
\end{aligned}
$$

The formal representation of the contributions of the semicircles by a δ-function gives the same result as the explicit calculation.

Fig. 2.15 Path of integration for the calculation of $G_0^{(-)}(r, r')$

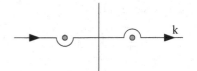

The Green's function $G_0^{(-)}(r, r')$ is calculated with the path of integration shown in Fig. 2.15. The contribution of the semicircles are evaluated by a path traversed in the opposite direction as compared to the calculation of $G_0^{(+)}$. The result is therefore

$$
\begin{aligned}
G_0^{(-)}(r, r') &= \frac{m_0}{2\pi \hbar^2 q} \left\{ \mathscr{P} \int_{-\infty}^{+\infty} \frac{dk}{\pi} \frac{k}{(k_0^2 - k^2)} \sin kq + i \sin k_0 q \right\} \\
&= -\frac{m_0}{2\pi \hbar^2 q} \left\{ \cos k_0 q - i \sin k_0 q \right\} \\
&= -\frac{m_0}{2\pi \hbar^2 q} e^{-i k_0 q}.
\end{aligned}
\tag{2.46}
$$

The definition of the Green's function with the boundary condition of standing waves

$$
G_0^{(s)}(r, r') = \frac{1}{2} \left[G_0^{(+)}(r, r') + G_0^{(-)}(r, r') \right]
$$

indicates that the contributions of the semicircles cancel, so that this Green's function is determined only by the principal value.

Literature in Chap. 2

1. B. A. Lippmann, Phys. Rev. Lett. **79**, p. 461 (1950)
2. R. P. Feynman, Phys. Rev. **76**, p. 749 (1949)

Elastic Scattering: Time-Dependent Formulation

The process of elastic scattering of two classical particles can be described as follows: The particles are initially so far apart, that there is no interaction between them. However, their trajectories are arranged that the particles interact with each other for a certain period of time, after which they separate again and move apart without further interaction. The scattering is elastic if the particles have no internal structure that can be excited at the cost of energy. If the interaction between the particles depends, as discussed in Sect. 1.1.2, only on the relative coordinate, the centre of mass motion of the two-particle system need not be discussed.

In the case of quantum particles (relativistic or nonrelativistic) a specification of particle trajectories is not possible. For a given initial state of the relative motion of the system $|\psi_i(t_i)\rangle$ the time development, including all quantum effects, is governed by time development operators (in the nonrelativistic case based on the Schrödinger equation) for the quantum states. The outcome of the collision experiment is analysed by projection of the final state $|\psi_f(t_f)\rangle$ onto plane wave states. In order to represent the time development of the states, the time variation of the distance between the classical particles is simulated by adiabatic switching (on and off) of the interaction. In addition, the times t_i and t_f are replaced by the limiting values $t_i = -\infty$ and $t_f = +\infty$. The transition from the initial state to the final state takes place in a symmetric time interval around the time $t = 0$ and does not involve a dependence on a time coordinate t.

In order to discuss the collision with the aid of a limiting process involving the time coordinates, one introduces two operators, the Møller operators $\hat{\Omega}_+$ and $\hat{\Omega}_-$. The operator $\hat{\Omega}_+$ describes the time development of the system from the time $t = -\infty$ to the time $t = 0$, the operator $\hat{\Omega}_-$ from $t = +\infty$ back to the time $t = 0$. With the Møller operators one constructs the operator for the S-matrix

$$\hat{S} = \hat{\Omega}_-^\dagger \hat{\Omega}_+ ,$$

which describes the complete collision process. The time-independent S-matrix operator is a central quantity of the time-dependent approach to the scattering theory. The critical question, that has to be answered, is the question whether the limiting process in time, indicated above, can be executed without problems.

The properties of the three operators $\hat{\Omega}_{\pm}$ and $\hat{S}$ are specified in this chapter on the basis of their definition as limiting values in time. It can then be shown that the S-matrix operator and the T-matrix operator are related, so that the stationary and time-dependent treatment of the collision processes yield equivalent results.

The remainder of the chapter deals with two aspects, which are more technical. A third operator for the discussion of scattering processes, the K-matrix operator, can be introduced. One of its properties is the fact, that K-matrix elements satisfy the optical theorem, even if the operator itself is only available approximately. In order to round off the discussion of potential scattering, the last point of this chapter is the partial wave expansion of the matrix elements of the operators $\hat{T}$, $\hat{S}$ and $\hat{K}$ for central and for more general potentials.

3.1   The Time Development Operator

The nonrelativistic, quantum mechanical initial value problem implies the solution of the time-dependent Schrödinger equation with a time-independent Hamiltonian[1] (compare Chap. 2)

$$i\hbar \frac{\partial}{\partial t} |\Psi(t)\rangle = \left[\hat{H}_0 + \hat{V} \right] |\Psi(t)\rangle .$$

The discussion of such initial value problems is conveniently based on the formulation of time-dependent quantum mechanics in specific representations.

- The *Schrödinger picture* (indexed with S), in which the time development of the states is determined by the time-independent Hamiltonian. All other operators are normally time-independent in this representation.
- The *interaction picture* (indexed with I for *interaction*), in which the time development of the states by the operator $\hat{H}_0$ is separated from the time development by the operator $\hat{V}$. Operators are necessarily time-dependent in this picture.

The two representations are connected by a unitary transformation. This implies that inner products of state vectors and other matrix elements are independent of the

---

[1] This is also a basic problem, that is addressed as part of the introduction to quantum mechanics. If the potential function is explicitly time-dependent—for example, in the case of the motion of interacting particles in an additional, oscillating electromagnetic field—then a more extensive consideration is required.

particular representation, i.e. a particular representation can be chosen freely. In the following, it is assumed (for simplicity) that the operator $\hat{H}_0$ is identical with the operator for the kinetic energy $\hat{T}$.

The course of an individual collision process can be described with the help of time development operators. In order to connect with the stationary formulation, one has to describe the process formally[2] from very early to very late times. The discussion of the limiting values in time, without an explicit analysis of the processes at small separations of the collision partners, finally leads, via the discussion of the Møller operators and the S-matrix, to the statement that the stationary and the time-dependent treatment of the scattering problem are equivalent.

3.1.1 The Schrödinger Picture

For a scattering problem, an initial state (initial time t_i)

$$|\psi_S(t_i)\rangle \equiv |\psi_{in}\rangle$$

is specified in the form of a wave packet for the relative motion with a certain localisation compatible with the uncertainty principle in space and momentum

$$|\psi_{in}\rangle = \int d^3k \, |k\rangle\langle k|\psi_{in}\rangle = \int d^3k \, \psi_{in}(k)|k\rangle.$$

This wave function can be expanded in terms of plane waves

$$\psi_{in}(r) = \langle r|\psi_{in}\rangle = \int d^3k \, \psi_{in}(k)\frac{e^{ik\cdot r}}{(2\pi)^{3/2}}.$$

Direct integration of the time-dependent Schrödinger equation over time

$$i\hbar\partial_t|\psi_S(t)\rangle = \hat{H}|\psi_S(t)\rangle = (\hat{H}_0 + \hat{V})|\psi_S(t)\rangle$$

leads to an equivalent integral equation for the time-dependent state, which has developed from $|\psi_{in}\rangle$

$$|\psi_S(t)\rangle = |\psi_{in}\rangle - \frac{i}{\hbar}\int_{t_i}^{t} dt' \hat{H}|\psi_S(t')\rangle. \tag{3.1}$$

The initial condition is incorporated in (3.1) One can describe the change of the state $|\psi_S(t)\rangle$ with time by introducing a *time development operator* $\hat{U}_S(t, t_i)$, which

---

[2] The interaction time can in reality be very short.

in the general case takes the form

$$|\psi_S(t)\rangle = \hat{U}_S(t, t_i)|\psi_S(t_i)\rangle. \tag{3.2}$$

For the scattering problem one writes

$$|\psi_S(t)\rangle = \hat{U}_S(t, t_i)|\psi_{in}\rangle$$

with initial value

$$\hat{U}_S(t_i, t_i) = \hat{1}.$$

The ansatz (3.2) yields instead of (3.1) an operator equation

$$\hat{U}_S(t, t_i) = \hat{1} - \frac{i}{\hbar} \int_{t_i}^t dt' \hat{H} \, \hat{U}_S(t', t_i), \tag{3.3}$$

which may act on any initial state. A formal solution of the integral equation (3.3) is obtained by iteration[3]

$$\hat{U}_S^{(1)}(t, t_i) = \hat{1} + \left(-\frac{i}{\hbar}\right) \hat{H}(t - t_i),$$

$$\hat{U}_S^{(2)}(t, t_i) = \hat{1} + \left(-\frac{i}{\hbar}\right) \hat{H}(t - t_i) + \frac{1}{2!}\left(-\frac{i}{\hbar}\right)^2 \hat{H}^2(t - t_i)^2,$$

$$\vdots$$

$$\hat{U}_S^{(n+1)}(t, t_i) = \hat{U}_S^{(n)}(t, t_i) + \frac{1}{(n+1)!}\left(-\frac{i}{\hbar}\right)^{n+1} \hat{H}^{n+1}(t - t_i)^{n+1},$$

$$\vdots$$

and resummation

$$\hat{U}_S(t, t_i) = \exp\left[-\frac{i}{\hbar}\hat{H}(t - t_i)\right]. \tag{3.4}$$

The formal solution (3.4) stands for the corresponding operator series. Only in simple cases is it possible to find a closed analytic expression for $\hat{U}_S(t, t_i)$.

---

[3] The simple form of (3.3) is only valid for a *time-independent* Hamiltonian.

The operator $\hat{U}_S(t, t_i)$ has the following properties:

- It is *unitary*. From (3.4) follows

$$\hat{U}_S^\dagger(t, t_i) = \exp\left[\frac{\mathrm{i}}{\hbar}\hat{H}(t - t_i)\right] = \exp\left[-\frac{\mathrm{i}}{\hbar}\hat{H}(t_i - t)\right]$$
$$= \hat{U}_S(t_i, t) = \hat{U}_S^{-1}(t, t_i), \tag{3.5}$$

provided the Hamiltonian $\hat{H}$ is Hermitian $\hat{H} = \hat{H}^\dagger$. The Hermitian adjoint of the time development operator corresponds to the inverse time development operator, it describes the time development in the inverse direction.
- It satisfies the multiplication theorem

$$\hat{U}_S(t, t_1)\,\hat{U}_S(t_1, t_i) = \hat{U}_S(t, t_i). \tag{3.6}$$

This follows from the general relation

$$e^{\hat{A}}e^{\hat{B}} = e^{\hat{A}+\hat{B}} \qquad \text{for} \qquad [\hat{A}, \hat{B}] = \hat{0}.$$

- With the statements (3.5) and (3.6) one can note the equations[4]

$$\hat{U}_S^\dagger(t, t_i)\hat{U}_S(t, t_i) = \hat{1} \quad \text{and} \quad \hat{U}_S(t, t_i)\hat{U}_S^\dagger(t, t_i) = \hat{1}. \tag{3.7}$$

The unitary operator $\hat{U}_S$ conserves the norm and inner products

$$\langle\phi|\hat{U}_S^\dagger(t, t_i)\hat{U}_S(t, t_i)|\chi\rangle = \langle\phi|\chi\rangle.$$

- The time development operator

$$\hat{U}_{S,0}(t, t_i) \equiv \hat{U}_0(t, t_i) = \exp\left[-\frac{\mathrm{i}}{\hbar}\hat{H}_0(t - t_i)\right] \tag{3.8}$$

describes free propagation, if the operator $\hat{H}_0$ represents the kinetic energy[5]

$$\hat{H}_0 \equiv \hat{T}.$$

---

[4] From $\hat{O}^\dagger\hat{O} = \hat{1}$ does not necessarily follow $\hat{O}\hat{O}^\dagger = \hat{1}$. An operator $\hat{O}$, with the property $\hat{O}^\dagger\hat{O} \neq \hat{O}\hat{O}^\dagger$ is not unitary.
[5] This statement is assumed for the remaining part of this chapter.

3.1.2   The Interaction Picture

The states and operators in the interaction picture are connected via a unitary transformation mediated by an operator $\hat{U}_0(t, t_0)$ with the states and operators in the Schrödinger representation. For the states one writes

$$|\psi_I(t)\rangle = \hat{U}_0^\dagger(t, t_0)|\psi_S(t)\rangle. \tag{3.9}$$

The states in the two pictures coincide at an arbitrary time $t = t_0$. For the transformation of operators in the two pictures follows: The Schrödinger equation in the interaction picture is

$$i\hbar \partial_t |\psi_I(t)\rangle = \hat{V}_I(t)|\psi_I(t)\rangle,$$

if the time-dependent interaction $\hat{V}_I(t)$ is defined by

$$\hat{V}_I(t) = \hat{U}_0^\dagger(t, t_0)\hat{V}_S\hat{U}_0(t, t_0).$$

The time development of the state $|\psi_I(t)\rangle$ is determined by the interaction $V_I(t)$ alone. If one differentiates this relation with respect to time, one finds

$$\partial_t \hat{V}_I(t) = \frac{i}{\hbar}\left[\hat{H}_0, \hat{V}_I(t)\right].$$

The time dependence of the interaction in the interaction picture is determined by the commutator of $\hat{V}_I(t)$ with $\hat{H}_0$.

Similarly, one has for any other operator

$$\hat{O}_I(t) = \hat{U}_0^\dagger(t, t_0)\hat{O}_S\hat{U}_0(t, t_0).$$

The time development of $\hat{O}_I(t)$ is governed by the same equation of motion

$$\partial_t \hat{O}_I(t) = \frac{i}{\hbar}[\hat{H}_0, \hat{O}_I(t)].$$

The free Hamiltonian has the same form in the two pictures.

In order to obtain an explicit representation of the time development operator $\hat{U}_I$ in the interaction picture

$$|\psi_I(t)\rangle = \hat{U}_I(t, t_1)|\psi_I(t_1)\rangle$$

one uses (with (3.2) and (3.9))

$$
\begin{aligned}
|\psi_I(t)\rangle &= \hat{U}_0^\dagger(t, t_0)|\psi_S(t)\rangle \\
&= \hat{U}_0^\dagger(t, t_0)\hat{U}_S(t, t_1)|\psi_S(t_1)\rangle \\
&= \hat{U}_0^\dagger(t, t_0)\hat{U}_S(t, t_1)\hat{U}_0(t_1, t_0)|\psi_I(t_1)\rangle
\end{aligned}
$$

and identifies

$$
\hat{U}_I(t, t_1) = \hat{U}_0^\dagger(t, t_0)\hat{U}_S(t, t_1)\hat{U}_0(t_1, t_0), \tag{3.10}
$$

so that

$$
\hat{U}_I(t_1, t_1) = \hat{1}.
$$

The operator $\hat{U}_I(t_2, t_1)$ has the same properties as the operator $\hat{U}_S(t_2, t_1)$. It is unitary, as

$$
\hat{U}_I^\dagger(t_2, t_1)\hat{U}_I(t_2, t_1) = \hat{U}_I(t_2, t_1)\hat{U}_I^\dagger(t_2, t_1) = \hat{1}, \tag{3.11}
$$

and it obeys the multiplication theorem

$$
\hat{U}_I(t_3, t_2)\hat{U}_I(t_2, t_1) = \hat{U}_I(t_3, t_1). \tag{3.12}
$$

The Schrödinger equation in the interaction picture corresponds to the integral equation

$$
\hat{U}_I(t, t_i) = \hat{1} - \frac{i}{\hbar}\int_{t_i}^t dt'\, \hat{V}_I(t')\hat{U}_I(t', t_i) \tag{3.13}
$$

for the operator $\hat{U}_I(t, t_i)$. The iteration of this integral equation is more complicated, since the interaction operators at different times do not commute

$$
\hat{V}_I(t_1)\hat{V}_I(t_2) \neq \hat{V}_I(t_2)\hat{V}_I(t_1).
$$

This implies that one has to observe the exact time sequence in each order of the iteration. The iteration of (3.13) leads for this reason to

$$
\hat{U}_I(t, t_i) = T\left[e^{-\frac{i}{\hbar}\int_{t_i}^t dt'\, \hat{V}_I(t')}\right], \tag{3.14}
$$

where the notation T characterises a time-ordered product of the operators in the expansion

$$
\mathrm{T}\left[e^{-\frac{i}{\hbar}\int_{t_i}^{t} dt' \hat{V}_I(t')}\right] = \hat{1} + \sum_{n=1}^{\infty} \left(\frac{1}{n!}\right)\left(-\frac{i}{\hbar}\right)^n \int_{t_i}^{t} dt_n \int_{t_i}^{t} dt_{n-1} \cdots
$$

$$
\cdots \int_{t_i}^{t} dt_1 \, \mathrm{T}\,[\hat{V}_I(t_n) \cdots \hat{V}_I(t_1)] + \ldots
$$

$$
= \hat{1} + \sum_{n=1}^{\infty} \left(-\frac{i}{\hbar}\right)^n \int_{t_i}^{t} dt_n \int_{t_i}^{t_n} dt_{n-1} \cdots
$$

$$
\cdots \int_{t_i}^{t_2} dt_1 [\hat{V}_I(t_n) \cdots \hat{V}_I(t_1)] + \ldots
$$

(3.15)

The explicit form of the time-ordered product is given in (3.15). The operators are arranged so that operators at earlier times in the operator chain are always to the right of operators at later times.

3.1.3   The Time Development of the Collision Process

In order to follow an (elastic) scattering process in detail, one should trace the time development from an initial time $t_i = t_{\mathrm{in}}$ at which the collision partners are separated, up to a point in time $t_f = t_{\mathrm{out}}$ at which the particles are separated again. A wave packet at the initial time will evolve up to the time of the closest approach of the collision partners and beyond under the influence of the mutual interaction. At the end of the experiment a final state is reached, which corresponds to a uniform, relative motion of the collision partners. In order that this scenario takes place with initial and final states, that involve only a uniform motion, one must assume that the interaction between the collision partners is limited to a finite domain in space. Then, initially, only free propagation takes place. The interaction is switched on with the approach of the particles and after the closest approach is switched off again if the separation has increased sufficiently. Propagation in the final stage is free again.

While the actual experiment might take place in a short time interval, the initial and final phases of the collision process can be extended arbitrarily in time if only free motion occurs. The process can be increased to a time interval from $t_i = -\infty$ until $t_f = +\infty$. In order to demonstrate that the time-dependent and the stationary formulations lead to the same results for quantities such as differential cross sections, one has to examine the extension of the interval, in which the interaction takes place in detail. The question must be answered as to whether— or under what conditions—the two limiting values $t \longrightarrow \pm\infty$ do exist. For this purpose, one can use the interaction picture. One chooses as time of the closest approach of the collision partners $t_0 = 0$. According to the discussion in Sect. 3.1.2 the form of the wave packet in the two pictures agrees at this time. In the interaction picture the free particle states $|\psi_I(t)\rangle$ with $t \leq t_{in}$ and with $t \geq t_{out}$ do not change with time, so that one can prove the existence of the two limiting values in the form

$$|\psi_S(0)\rangle \equiv |\psi_I(0)\rangle = \lim_{t \to -\infty} \hat{U}_I(0, t)|\psi_{in}\rangle = \lim_{t \to -\infty} \hat{U}_I^\dagger(t, 0)|\psi_{in}\rangle \quad (3.16)$$

for the incoming part and

$$|\psi_S(0)\rangle \equiv |\psi_I(0)\rangle = \lim_{t \to +\infty} \hat{U}_I^\dagger(t, 0)|\psi_{out}\rangle \quad (3.17)$$

for the second half of the scattering process. The existence of the limiting values can be demonstrated (details in Sect. 3.6.1) on the basis of the integral equation (3.13)

$$\hat{U}_I(t, t_i) = \hat{1} - \frac{i}{\hbar} \int_{t_i}^t dt' \hat{V}_I(t') \hat{U}_I(t', t_i).$$

This discussion can, on one hand, be carried through with the specification of an explicit wave packet, whereby one assumes that the exact form of the packet does not play a significant role for the result of the limiting process. This option is discussed in Sect. 3.6.2. Alternatively, one can try to use the integral equation directly. As this equation is not well defined for $t_i \to -\infty$, regularisation is necessary. This requires the following steps:

- Substitute the interaction $\hat{V}_I(t)$ by

$$\hat{V}_I(t) \Longrightarrow \hat{V}_\epsilon(t) = e^{-\frac{\epsilon}{\hbar}|t|} \hat{V}_I(t).$$

  Here ϵ is assumed to be sufficiently small and larger than zero $\epsilon > 0$.
- Solve or discuss the integral equation with $\hat{V}_\epsilon(t)$.
- Determine the limiting value of this solution for $\epsilon \to 0$.

This explicit prescription, which is called *adiabatic switching*, assumes the switching (on and off) of the interaction as a function of the distance of the collision partners. This is in reality not possible. The *regularisation* with

$$\hat{V}_\epsilon(t) \overset{t \to \pm\infty}{\longrightarrow} 0 \quad \text{for} \quad \epsilon > 0 \quad \text{and} \quad \hat{V}_\epsilon(0) = \hat{V}_I(0)$$

simulates an adiabatic switching of the interaction $\hat{V}_I(t)$ which does not distort the correct time flow, if the switching parameter ϵ is small enough. This statement is examined in more detail in Sect. 3.6.3. Combining the results of the two possible approaches, one can state:

The limits (3.16) and (3.17) exist, if the potential energy is continuous with the exception of a finite number of points and if it satifies the conditions (compare

Sect. 1.1.2)

$$v(r) \xrightarrow{r \to 0} \frac{1}{r^{3/2-\delta}} \quad \text{and} \quad v(r) \xrightarrow{r \to \infty} \frac{1}{r^{3/2+\delta}} \quad (\delta > 0).$$

The existence of the time limits allows the definition of two Møller operators,[6] two central operators of scattering theory

$$\hat{\Omega}_{\pm} = \lim_{t \to \mp \infty} \hat{U}_S^{\dagger}(t, 0)\hat{U}_0(t, 0) = \lim_{t \to \mp \infty} \hat{U}_I^{\dagger}(t, 0) \tag{3.18}$$

with the properties

$$|\psi_S(0)\rangle = \hat{\Omega}_+|\psi_{\text{in}}\rangle \quad \text{and} \quad |\psi_S(0)\rangle = \hat{\Omega}_-|\psi_{\text{out}}\rangle, \tag{3.19}$$

which correspond to the limiting values in (3.16) and (3.17).

3.1.4 The Møller Operators

The relations (3.18) and (3.19) can be used to obtain additional and useful statements about the Møller operators:

- The requirement of norm conservation of the states is satisfied as one finds

$$\langle \psi_S(0)|\psi_S(0)\rangle = \langle \psi_{\text{in}}|\hat{\Omega}_+^{\dagger}\hat{\Omega}_+|\psi_{\text{in}}\rangle = \langle \psi_{\text{out}}|\hat{\Omega}_-^{\dagger}\hat{\Omega}_-|\psi_{\text{out}}\rangle = 1,$$

  if the relation

$$\hat{\Omega}_+^{\dagger}\hat{\Omega}_+ = \hat{\Omega}_-^{\dagger}\hat{\Omega}_- = \hat{1} \tag{3.20}$$

  holds.
- Inversion of the second equation in (3.19) and insertion into the first yields

$$|\psi_{\text{out}}\rangle = \hat{\Omega}_-^{\dagger}|\psi_S(0)\rangle = \hat{\Omega}_-^{\dagger}\hat{\Omega}_+|\psi_{\text{in}}\rangle. \tag{3.21}$$

This implies that an operator can be obtained by combining the two Møller operators, which transforms the initial state directly into the final state. The initial state is prepared as a free state, and after the scattering process the final state is observed as a free state. The operator product

$$\hat{S} = \hat{\Omega}_-^{\dagger}\hat{\Omega}_+, \tag{3.22}$$

---

[6] C. Møller, Danske Videnskab. Selskab, Mat-fys. Medd. **23**, p. 1 (1948).

which contains all the necessary information about the scattering process, is called the *S-matrix operator*.[7] In view of its central importance, it will be analysed more closely in the following section.

- For finite values of time τ one finds

$$\lim_{t \to \pm\infty} e^{\frac{i}{\hbar}\hat{H}\tau} \left[e^{\frac{i}{\hbar}\hat{H}t} e^{-\frac{i}{\hbar}\hat{H}_0 t} \right] e^{-\frac{i}{\hbar}\hat{H}_0\tau} = \lim_{t' \to \pm\infty} e^{\frac{i}{\hbar}\hat{H}t'} e^{-\frac{i}{\hbar}\hat{H}_0 t'},$$

where $t' = t + \tau$ has been used. On both sides of this equation one recognises Møller operators, so that one can write

$$e^{\frac{i}{\hbar}\hat{H}\tau}\hat{\Omega}_\pm e^{-\frac{i}{\hbar}\hat{H}_0\tau} = \hat{\Omega}_\pm \quad \text{or} \quad e^{\frac{i}{\hbar}\hat{H}\tau}\hat{\Omega}_\pm = \hat{\Omega}_\pm e^{\frac{i}{\hbar}\hat{H}_0\tau}.$$

In first order the expansion of the exponential operators gives

$$\hat{H}\hat{\Omega}_\pm = \hat{\Omega}_\pm \hat{H}_0. \tag{3.23}$$

If one uses Eq. (3.20), one obtains the relation

$$\hat{\Omega}_\pm^\dagger \hat{H}\hat{\Omega}_\pm = \hat{H}_0. \tag{3.24}$$

A similarity transformation of the full Hamiltonian with the Møller operators leads to the free Hamiltonian. This shows that the Møller operators can not be unitary in general, since (3.24) would imply that the Hamiltonians $\hat{H}$ and $\hat{H}_0$ have the same spectrum. However, this is only the case if the spectrum of $\hat{H}$ does not contain bound states. In inversion of this statement one can state, that the Møller operators are unitary, if the spectrum of $\hat{H}$ does not contain bound states.
- Møller operators are isometric operators. The difference with respect to unitary operators is: unitary operators map the entire solution space of a problem with a Hamiltonian $\hat{H}$ onto itself, isometric operators do not (see Sect. 3.3.1).

3.2 The S-Matrix

The S-matrix operator

$$\hat{S} = \hat{\Omega}_-^\dagger \hat{\Omega}_+$$

results from a limiting process ($t \to \pm\infty$) within the framework of the time-dependent treatment of the potential scattering problem. For this reason it links the discussion of the scattering problem on the basis of a stationary treatment with the

---

[7] This concept was first introduced by J. A. Wheeler, Phys. Rev. **52**, p. 1107 (1937) and later discussed in a more general form and extended by W. Heisenberg, Z. Phys. Rev. **120**, p. 513 (1943).

actual time-dependent process of the collision. The question, that should be posed at this stage, is: How can the S-matrix operator be related to the elements of stationary scattering theory that have been discussed in the preceeding chapters?

A direct statement about S-matrix elements is: only on-shell S-matrix elements exist. This is a consequence of Eq. (3.24), which is used in the following brief chain of transformations

$$\hat{S}\hat{H}_0 = \hat{\Omega}_-^\dagger \hat{\Omega}_+ \hat{H}_0 = \hat{\Omega}_-^\dagger \hat{H} \hat{\Omega}_+ = \hat{H}_0 \hat{\Omega}_-^\dagger \hat{\Omega}_+ = \hat{H}_0 \hat{S}.$$

The S-matrix operator commutes with the free Hamiltonian

$$[\hat{H}_0, \hat{S}] = 0, \tag{3.25}$$

so that

$$\langle k | [\hat{H}_0, \hat{S}] | k' \rangle = (E(k) - E(k')) \langle k | \hat{S} | k' \rangle = 0$$

follows immediately. The S-matrix elements vanish if the energy $E(k)$ is not equal to the energy $E(k')$.

3.2.1   Adiabatic Switching, Formal Considerations

A formal solution of the integral equation (3.13) by iteration is possible, but the subsequent extraction of the S-matrix, which describes the transition from $t_i \rightarrow -\infty$ to $t_f \rightarrow +\infty$, leads to difficulties. To circumvent these difficulties, a regularisation procedure is necessary. For the discussion of the integral equation (3.13) and ultimately the extraction of the S-matrix, one also uses the adiabatic switching introduced in Sect. 3.1.3 with

$$\hat{V}_I(t) \Longrightarrow \hat{V}_\epsilon(t) = e^{-\frac{\epsilon}{\hbar}|t|} \hat{V}_I(t).$$

As the parameter ϵ is assumed to be $\epsilon > 0$ and sufficiently small, all terms of the iterative solution can be obtained for the function $\hat{V}_\epsilon(t)$. Therefore the S-matrix elements can be extracted. In the last step the limit $\epsilon \rightarrow 0$ is taken, which leads back to the original interaction.

The procedure can be understood in the following fashion: Without regularisation it is not possible to express the switching of the interaction as a function of the distance of the colliding partners, or the distance of the colliding particle from the *target*. The regularisation simulates because of

$$\hat{V}_\epsilon(t) \xrightarrow{t \rightarrow \pm\infty} 0 \quad \text{for} \quad \epsilon > 0 \quad \text{and} \quad \hat{V}_\epsilon(0) = \hat{V}_I(0)$$

an adiabatic switching of the interaction $\hat{V}_I(t)$, which does not distort the correct flow of time, if the switching parameter ϵ is sufficiently small.

Iteration of the integral equation (3.13) after the substitution $\hat{V}_I \rightarrow V_\epsilon$ gives

$$\hat{U}_\epsilon(t, t_i) = \hat{1} - \frac{i}{\hbar} \int_{t_i}^{t} dt_1 \hat{V}_\epsilon(t_1) + \left(\frac{i}{\hbar}\right)^2 \int_{t_i}^{t} dt_1 \hat{V}_\epsilon(t_1) \int_{t_i}^{t_1} dt_2 \hat{V}_\epsilon(t_2) + \ldots$$

or in summary

$$\hat{U}_\epsilon(t, t_i) = \sum_{n=0}^{\infty} \left(-\frac{i}{\hbar}\right)^n \int_{t_i}^{t} dt_1 \hat{V}_\epsilon(t_1) \int_{t_i}^{t_1} dt_2 \hat{V}_\epsilon(t_2) \cdots$$
$$\cdots \int_{t_i}^{t_{n-1}} dt_n \hat{V}_\epsilon(t_n).$$

The S-matrix operator is defined according to (3.18) and (3.22) as

$$\hat{S} = \hat{\Omega}_-^\dagger \hat{\Omega}_+ = \lim_{\substack{t_1 \to \infty \\ t_2 \to -\infty}} \left[\hat{U}_0^\dagger(t_1, 0) \hat{U}_S(t_1, 0) \hat{U}_S^\dagger(t_2, 0) \hat{U}_0(t_2, 0) \right]$$

$$= \lim_{\substack{t_1 \to \infty \\ t_2 \to -\infty}} \hat{A}(t_1, t_2).$$

The products of time development operators correspond exactly to the time development operator in the interaction picture (because of (3.10))

$$\hat{U}_I(t, 0) = \left[\hat{U}_0^\dagger(t, 0) \hat{U}_S(t, 0) \right],$$

so that

$$\hat{A}(t_1, t_2) = \hat{U}_I(t_1, 0) \hat{U}_I^\dagger(t_2, 0).$$

After applying the conjugation and the multiplication rules one obtains

$$\hat{A}(t_1, t_2) = \hat{U}_I(t_1, t_2). \tag{3.26}$$

At this point regularisation comes into play. One computes the S-matrix operator, or ultimately the S-matrix elements, via the replacement of $\hat{U}_I$ by $\hat{U}_\epsilon$ and the limiting process $\epsilon \to 0$

$$\hat{S} = \lim_{\epsilon \to 0} \hat{S}_\epsilon = \lim_{\epsilon \to 0} \left\{ \sum_{n=0}^{\infty} \left(-\frac{i}{\hbar} \right)^n \int_{-\infty}^{\infty} dt_1 \hat{V}_\epsilon(t_1) \int_{-\infty}^{t_1} dt_2 \hat{V}_\epsilon(t_2) \cdots \right.$$

$$\left. \cdots \int_{-\infty}^{t_{n-1}} dt_n \hat{V}_\epsilon(t_n) \right\}.$$

The limiting process $\epsilon \to 0$ is to be carried out after evaluation of the integrals. The corresponding S-matrix elements with respect to plane wave states

$$\langle k' | \hat{S} | k \rangle = \lim_{\epsilon \to 0} \langle k' | \hat{S}_\epsilon | k \rangle$$

can only be evaluated order by order. One obtains

- to zeroth order

$$\langle k' | \hat{S} | k \rangle_{(0)} = \delta(k - k').$$

The particles pass each other without interaction. The relative momentum does not change.
- For the first order, the starting point is

$$\langle k' | \hat{S} | k \rangle_{(1)} = - \lim_{\epsilon \to 0} \left\{ \frac{i}{\hbar} \int_{-\infty}^{\infty} dt_1 \, e^{\frac{\epsilon}{\hbar} |t_1|} \, \langle k' | e^{\frac{i}{\hbar} \hat{H}_0(t_1)} \hat{V} e^{-\frac{i}{\hbar} \hat{H}_0(t_1)} | k \rangle \right\}$$

$$= - \lim_{\epsilon \to 0} \left\{ \frac{i}{\hbar} \int_{-\infty}^{\infty} dt_1 \, \exp\left[-\frac{i}{\hbar} \{ (E(k) - E(k'))t_1 - i\epsilon|t_1| \} \right] \right\}$$

$$\times \, \langle k' | \hat{V} | k \rangle.$$

In this expression, the limiting process and the integration can be interchanged (see Sect. 3.6.4), so that one is left with a representation of the δ function

$$\langle k' | \hat{S} | k \rangle_{(1)} = -2\pi i \, \delta(E(k) - E(k')) \langle k' | \hat{V} | k \rangle.$$

Here both the on-shell behaviour and the (expected) proportionality to the potential matrix element find their expression.
- In second order, the matrix element

$$\langle k' | \hat{S} | k \rangle_{(2)} = \left(\frac{-i}{\hbar} \right)^2 \lim_{\epsilon \to 0} \left\{ \int d^3k'' \int_{-\infty}^{\infty} dt_1 \, \langle k' | e^{\frac{i}{\hbar} \hat{H}_0(t_1)} \hat{V} e^{-\frac{i}{\hbar} \hat{H}_0(t_1)} | k'' \rangle \right.$$

$$\left. \times e^{-\frac{\epsilon}{\hbar} |t_1|} \int_{-\infty}^{\infty} dt_2 \, e^{\frac{-\epsilon}{\hbar} |t_2|} \, \langle k'' | e^{\frac{i}{\hbar} \hat{H}_0(t_2)} \hat{V} e^{-\frac{i}{\hbar} \hat{H}_0(t_2)} | k \rangle \right\}$$

has to be calculated. The explicit evaluation is somewhat more tedious (see Sect. 3.6.4). One finds in the end

$$\langle k' \hat{S} | k \rangle_{(2)} = -2\pi i\, \delta(E(k) - E(k')) \langle k' | \hat{V} \frac{1}{(E(k) - \hat{H}_0 + i\epsilon)} \hat{V} | k \rangle.$$

Again, the on-shell condition appears automatically. In addition, one can see the effect of the regularisation. It corresponds to the exclusion or the inclusion of the poles, which is used to implement the boundary conditions in the stationary treatment of the scattering problem.

3.2.2   The Relation Between the S- and the T-Matrix

On the basis of the perturbative evaluation one finds in lowest order the corresponding expansion of the T-matrix elements (2.34), with the exception of the energy-conserving factor. With a little more effort (Sect. 3.6.5), one can show that the two sets of matrix elements are related by

$$\langle k' | \hat{S} | k \rangle = \delta(k - k') - 2\pi i\, \delta(E(k) - E(k')) \langle k' | \hat{T} | k \rangle. \tag{3.27}$$

The corresponding operator form is

$$\hat{S} = \hat{1} - 2\pi i\, \delta(E_{\text{in}} - \hat{H}_0)\, \hat{T} = \hat{1} - 2\pi i\, \hat{T}\, \delta(E_{\text{out}} - \hat{H}_0). \tag{3.28}$$

The S-matrix elements contain the same information as the on-shell T-matrix elements. This statement supports the claim that the time-dependent formulation of the scattering theory as an initial value problem and the stationary formulation as a boundary value problem (with boundary conditions according to the extended Huygens' principle) are equivalent in the sense of the relation (3.27). With either formulation, one can compute the measured quantities, the differential or the total cross sections.

Additional statements resulting from the close connection of the two formulations are assembled in the next section.

3.3   Statements About the S-Matrix and the Møller Operators

The Lippmann-Schwinger equation (2.23)

$$| \psi_{k'}^{(-)} \rangle = | k' \rangle + \frac{1}{(E(k') - \hat{H} - i\epsilon)} \hat{V} | k' \rangle$$

is the starting point for an alternative representation of the S-matrix elements, which leads to a number of useful relations. To this end, one writes the Green's function with the Dirac identity (2.25)

$$\frac{1}{(E(k') - \hat{H} - i\epsilon)} = \frac{1}{(E(k') - \hat{H} + i\epsilon)} + 2\pi i\,\delta(E(k') - \hat{H})$$

and obtains

$$|\psi_{k'}^{(-)}\rangle = \left\{|k'\rangle + \frac{1}{(E(k') - \hat{H} + i\epsilon)}\hat{V}|k'\rangle\right\} + 2\pi i\,\delta(E(k') - \hat{H})\hat{V}|k'\rangle$$
$$= |\psi_{k'}^{(+)}\rangle + 2\pi i\,\delta(E(k') - \hat{H})\hat{V}|k'\rangle.$$

The calculation of the matrix element $\langle\psi_{k'}^{(-)}|\psi_k^{(+)}\rangle$ via

$$\langle\psi_{k'}^{(-)}|\psi_k^{(+)}\rangle = \langle\psi_{k'}^{(+)}|\psi_k^{(+)}\rangle - 2\pi i\,\langle k'|\hat{V}\delta(E(k') - \hat{H})|\psi_k^{(+)}\rangle$$
$$= \delta(k - k') - 2\pi i\,\delta(E(k') - E(k))\langle k'|\hat{V}|\psi_k^{(+)}\rangle$$

then results in the representation[8]

$$\langle\psi_{k'}^{(-)}|\psi_k^{(+)}\rangle = \langle k'|\hat{S}|k\rangle \tag{3.29}$$

because of (2.26) and (3.27).

3.3.1   Additional Properties of the Møller Operators

Equation (3.29) allows additional statements about the Møller operators, for example:

- It implies the representation

$$\hat{\Omega}_+ = \int d^3k\,|\psi_k^{(+)}\rangle\langle k|,$$
$$\hat{\Omega}_- = \int d^3k\,|\psi_k^{(-)}\rangle\langle k|, \tag{3.30}$$

  because the relation

$$\hat{S} = \hat{\Omega}_-^\dagger\hat{\Omega}_+ = \int d^3k_1\,d^3k_2\,|k_1\rangle\langle\psi_{k_1}^{(-)}|\psi_{k_2}^{(+)}\rangle\langle k_2|$$

---

[8] Only scattering states with the same boundary conditions are orthogonal (details in Sect. 3.6.6).

follows and thus

$$\langle k' | \hat{S} | k \rangle = \langle \psi_{k'}^{(-)} | \psi_k^{(+)} \rangle.$$

The representation (3.30) of the Møller operators represents an interface between the stationary (right-hand side) and the time-dependent (left-hand side) formulations of the potential scattering problem.

- For the Møller operators themselves, one finds with (3.30) (see also Sect. 3.6.3)

$$\hat{\Omega}_+ |k\rangle = |\psi_k^{(+)}\rangle,$$
$$\hat{\Omega}_- |k\rangle = |\psi_k^{(-)}\rangle. \tag{3.31}$$

This result explains the notation with the indices $\pm$. The application of the Møller operators to plane waves produces exact scattering states with corresponding boundary conditions.[9]

- The product $\hat{\Omega}_+^\dagger \hat{\Omega}_+$ corresponds to the unit operator. One begins with

$$\hat{\Omega}_+^\dagger \hat{\Omega}_+ = \int d^3k_1 d^3k_2 \, |k_1\rangle \langle \psi_{k_1}^{(+)} | \psi_{k_2}^{(+)} \rangle \langle k_2 |.$$

As the scattering states with the same boundary conditions are orthogonal and the set of plane wave states is complete, it follows that

$$\hat{\Omega}_+^\dagger \hat{\Omega}_+ = \hat{1}. \tag{3.32}$$

- The product $\hat{\Omega}_+ \hat{\Omega}_+^\dagger$ does not correspond to the unit operator, if the potential, which is responsible for the scattering also allows bound states $|n\rangle$. The corresponding argument is

$$\hat{\Omega}_+ \hat{\Omega}_+^\dagger = \int d^3k_1 d^3k_2 \, |\psi_{k_1}^{(+)}\rangle \langle k_1 | k_2 \rangle \langle \psi_{k_2}^{(+)} |$$
$$= \int d^3k_1 \, |\psi_{k_1}^{(+)}\rangle \langle \psi_{k_1}^{(+)} | = \hat{1} - \sum_n |n\rangle \langle n|. \tag{3.33}$$

This result shows again, that the Møller operators are unitary if and only if the Hamiltonian $\hat{H} = \hat{H}_0 + \hat{V}$ has no bound states. An operator with the property

$$\hat{O}^\dagger \hat{O} = \hat{1} \quad \text{but} \quad \hat{O}\hat{O}^\dagger \neq \hat{1}$$

---

[9] Compare also Sect. 3.6.3.

is, as already mentioned above, called *isometric*. The product $\hat{\Omega}_+\hat{\Omega}_+^\dagger$ represents an operator which projects onto the scattering states of the Hamiltonian $\hat{H}$ because the statement

$$\hat{\Omega}_+\hat{\Omega}_+^\dagger|\psi\rangle = |\psi\rangle - \sum_{n,n'} c_n |n'\rangle\langle n'|n\rangle = \int d^3k\, c(k)|\psi_k^{(+)}\rangle$$

holds for any state

$$|\psi\rangle = \int d^3k\, c(k)|\psi_k^{(+)}\rangle + \sum_n c_n|n\rangle.$$

- The first of the relations (3.31) can be converted into an integral equation for the Møller operator $\hat{\Omega}_+$ with the help of the Lippmann-Schwinger equation (2.13) or (2.22). The result is

$$\hat{\Omega}_+|k\rangle = \left[\hat{1} + \hat{G}_0^{(+)}(E_0)\hat{V}\hat{\Omega}_+\right]|k\rangle$$

or

$$= \left[\hat{1} + \hat{G}^{(+)}(E_0)\hat{V}\right]|k\rangle.$$

Formally, one writes

$$\hat{\Omega}_+ = \hat{1} + \hat{G}_0^{(+)}(E_0)\hat{V}\hat{\Omega}_+ = \hat{1} + \hat{G}^{(+)}(E_0)\hat{V}. \tag{3.34}$$

Using the definition of the Møller operators in terms of time development operators (3.18), one also recognises a representation of the scattering states by a limiting value in time

$$|\psi_k^{(+)}\rangle = \lim_{t\to-\infty} \hat{U}_I(0,t)|k\rangle. \tag{3.35}$$

- Statements, that correspond to (3.34) and (3.35), are also valid for the operator $\hat{\Omega}_-$.

3.3.2 Properties of the S-Matrix

A distinctive property of the S-matrix is its unitarity, which is expressed by

$$\langle k'|\hat{S}\hat{S}^\dagger|k\rangle = \langle k'|\hat{S}^\dagger\hat{S}|k\rangle = \delta(k - k'). \tag{3.36}$$

For the proof one uses the definition (3.22) and the properties (3.30) to (3.33) of the Møller operators. One finds

$$\langle k'|\hat{S}\hat{S}^\dagger|k\rangle = \langle k'|\hat{\Omega}^\dagger_-\hat{\Omega}_+\hat{\Omega}^\dagger_+\hat{\Omega}_-|k\rangle = \langle\psi^{(-)}_{k'}|\hat{\Omega}_+\hat{\Omega}^\dagger_+|\psi^{(-)}_k\rangle$$
$$= \langle\psi^{(-)}_{k'}|\{\hat{1} - \sum_n |n\rangle\langle n|\}|\psi^{(-)}_k\rangle = \delta(k - k')$$

as well as

$$\langle k'|\hat{S}^\dagger\hat{S}|k\rangle = \langle k'|\hat{\Omega}^\dagger_+\hat{\Omega}_-\hat{\Omega}^\dagger_-\hat{\Omega}_+|k\rangle = \langle\psi^{(+)}_{k'}|\hat{\Omega}_-\hat{\Omega}^\dagger_-|\psi^{(+)}_k\rangle$$
$$= \langle\psi^{(+)}_{k'}|\{\hat{1} - \sum_n |n\rangle\langle n|\}|\psi^{(+)}_k\rangle = \delta(k - k').$$

The last computational step follows in both cases from the orthogonality of the bound and the scattering states.

The unitarity of the S-matrix expresses the optical theorem. From the relation (3.28)

$$\hat{S} = \hat{1} - 2\pi i\,\delta(E - \hat{H}_0)\hat{T}$$

follows

$$\hat{S}^\dagger\hat{S} = \hat{1} + 2\pi i\hat{T}^\dagger\delta(E - \hat{H}_0) - 2\pi i\delta(E - \hat{H}_0)\hat{T}$$
$$- (2\pi i)^2\hat{T}^\dagger\delta(E - \hat{H}_0)\delta(E - \hat{H}_0)\hat{T} = \hat{1}.$$

The on-shell matrix element of this operator equation (compare with the discussion of the T-matrix in Sect. 2, in particular Eq. (2.42)) is

$$\langle k|\hat{T}^\dagger - \hat{T}|k'\rangle = 2\pi i\langle k|\hat{T}^\dagger\delta(E - \hat{H}_0)\hat{T}|k'\rangle.$$

3.4 The K-Matrix

In many cases it is only possible to obtain an approximate solution of the Lippmann-Schwinger equations. These solutions do not necessarily satisfy the optical theorem. The question whether one can find a formulation of the scattering problem such that this theorem is satisfied in every approximation is answered in the affirmative by the introduction of the *K-matrix*. The formal definition of this third variant of a scattering matrix is based on the relation

$$\hat{S} = (\hat{1} - i\pi\,\delta(E_0 - \hat{H}_0)\hat{K})\frac{1}{(\hat{1} + i\pi\,\delta(E_0 - \hat{H}_0)\hat{K})}. \tag{3.37}$$

E_0 is again the given initial energy. The operators

$$\hat{A} = (\hat{1} - i\pi\delta(E_0 - \hat{H}_0)\hat{K}) \quad \text{and} \quad \hat{B} = (\hat{1} + i\pi\delta(E_0 - \hat{H}_0)\hat{K})$$

commute, as one has

$$[\hat{A}, \hat{B}] = (\hat{1} - i\hat{C})(\hat{1} + i\hat{C}) - (\hat{1} + i\hat{C})(\hat{1} - i\hat{C}) = \hat{0}.$$

Then $[\hat{A}, \hat{B}^{-1}] = \hat{0}$ is also valid, as multiplication of this commutator from the right and from the left with $\hat{B}$ yields $[\hat{A}, \hat{B}] = \hat{0}$. Equation (3.37), which defines the K-matrix, can therefore be extended to

$$\hat{S} = \hat{A}\,\hat{B}^{-1} = \hat{B}^{-1}\,\hat{A}.$$

From the unitarity condition for the S-matrix one then extracts the statement that the operator $\hat{K}$ must be Hermitian. For a proof of this statement one considers e.g.

$$\hat{S}^\dagger\hat{S} = \hat{A}^\dagger(\hat{B}^{-1})^\dagger\hat{A}\hat{B}^{-1} = (\hat{1} + i\hat{C}^\dagger)\frac{1}{(\hat{1} - i\hat{C}^\dagger)}(\hat{1} - i\hat{C})\frac{1}{(\hat{1} + i\hat{C})} \overset{!}{=} \hat{1}.$$

This requirement is satisfied, if $\hat{C}^\dagger = \hat{C}$, respectively if

$$\hat{K}^\dagger = \hat{K}. \tag{3.38}$$

A connection between the T-matrix and the K-matrix, in other words a connection between the K-matrix and the directly experimentally accessible quantity, is obtained from Eqs. (3.28) and (3.37)

$$\hat{S} = [\hat{1} - 2\pi i\delta(E_0 - \hat{H}_0)\hat{T}] = \frac{1}{(\hat{1} + i\pi\delta(E_0 - \hat{H}_0)\hat{K})}(\hat{1} - i\pi\delta(E_0 - \hat{H}_0)\hat{K})$$

via the steps: Multiply the equation from the left by

$$(\hat{1} + i\pi\delta(E_0 - \hat{H}_0)\hat{K})$$

and sort in the form

$$2\pi i\delta(E_0 - \hat{H}_0)\left\{\hat{K} - \hat{T} - i\pi\hat{K}\delta(E_0 - \hat{H}_0)\hat{T}\right\} = \hat{0}.$$

The expression in the curly brackets must vanish. The relation

$$\hat{T} = \hat{K} - i\pi\hat{K}\delta(E_0 - \hat{H}_0)\hat{T} \tag{3.39}$$

is known as *Heitler's damping equation*. If one starts with the relation $\hat{S} = \hat{A}\hat{B}^{-1}$, one finds with analogous steps the alternative form

$$\hat{T} = \hat{K} - i\pi\hat{T}\delta(E_0 - \hat{H}_0)\hat{K}. \tag{3.40}$$

From (3.40) one can, with a few computational steps, obtain an equation from which the K-matrix elements can be determined directly. For this purpose one inserts on both sides the Lippmann-Schwinger equation (2.28) for the T-matrix

$$\hat{V} + \hat{V}\hat{G}_0^{(+)}\hat{T} = \hat{K} - i\pi\hat{V}\delta(E_0 - \hat{H}_0)\hat{K} - i\pi\hat{V}\hat{G}_0^{(+)}\hat{T}\delta(E_0 - \hat{H}_0)\hat{K}.$$

The last term on the right-hand side can be rewritten with (3.40), so that one obtains

$$\hat{V} + \hat{V}\hat{G}_0^{(+)}\hat{T} = \hat{K} - i\pi\hat{V}\delta(E_0 - \hat{H}_0)\hat{K} + \hat{V}\hat{G}_0^{(+)}(\hat{T} - \hat{K}),$$

or, after sorting

$$\hat{K} = \hat{V} + \hat{V}\left[\frac{1}{(E_0 - \hat{H}_0 + i\epsilon)} + i\pi\delta(E_0 - \hat{H}_0)\right]\hat{K}.$$

The expression in the square brackets is the Green's function (2.18) for a boundary condition with standing waves

$$\hat{G}_0^{(s)} = \mathscr{P}\left(\frac{1}{(E_0 - \hat{H}_0)}\right),$$

so that the integral equation for the calculation of the K-matrix(elements) can be written in the form

$$\hat{K} = \hat{V} + \hat{V}\hat{G}_0^{(s)}\hat{K}. \tag{3.41}$$

The integral equation (3.41) differs from the integral equation for the T-matrix (2.28) only by the presence of the Green's function $\hat{G}_0^{(s)}$ instead of the Green's function $\hat{G}_0^{(+)}$. The matrix representation of $\hat{G}_0^{(s)}$ is real

$$\langle r|\hat{G}_0^{(s)}|r'\rangle = -\frac{m_0}{2\pi\hbar^2}\frac{\cos k|r - r'|}{|r - r'|},$$

so that the condition $\hat{K} = \hat{K}^{\dagger}$ is satisfied according to (3.41) for any exact and for any approximate solution of this equation. Thus, for the treatment of potential scattering, the following alternative scheme can be suggested:

- In the first step, one solves the integral equation (3.41) exactly, or if this is not feasible, approximately. In any case, the solution satisfies the basic condition of unitarity of the S-matrix as long as $\hat{K}$ is Hermitian.
- In the next step one computes the T-matrix elements with the damping equation (3.40). As a result of the presence of the δ-function, this is an on-shell equation. One only needs on-shell K-matrix elements in order to compute on-shell T-matrix elements. The optical theorem is satisfied for these matrix elements.

This procedure does, however, not guarantee that one obtains a useful solution for each approximation. One can satisfy the unitarity condition without having an acceptable solution. However, normally, satisfying this basic condition for a given approximation leads to an improvement of the result.

3.5   Partial Wave Expansions for the T-, S- and K-Matrices

A partial wave expansion, as described in Sect. 1.2 for the scattering amplitude, is also possible for the three scattering matrices. The simplest case with a spherically symmetric scattering potential $v(\boldsymbol{r}) = v(r) \longrightarrow \hat{V}_c$ is characterised by

$$[\hat{V}_c, \hat{\boldsymbol{l}}] = \hat{\boldsymbol{0}}.$$

The potential operator $\hat{V}_c$ commutes with the operators for the components $\hat{\boldsymbol{l}} = \{l_x, l_y, l_z\}$ of the orbital angular momentum. The vector $\hat{\boldsymbol{l}}$ represents, depending on the situation, either the orbital angular momentum of a particle or the relative angular momentum of two particles scattering from each other.

3.5.1   Central Potentials

In the momentum representation the matrix elements of a local potential are given as

$$\langle \boldsymbol{k}' | \hat{V}_c | \boldsymbol{k} \rangle = \int \mathrm{d}^3 r \, \langle \boldsymbol{k}' | \boldsymbol{r} \rangle v(r) \langle \boldsymbol{r} | \boldsymbol{k} \rangle.$$

If one inserts in this equation the expansion of the plane waves in terms of spherical harmonics.[10]

$$\langle r|k \rangle = \left(\frac{2}{\pi}\right)^{1/2} \sum_{lm} (i)^l j_l(kr) Y_{lm}^*(\Omega_r) Y_{lm}(\Omega_k), \tag{3.42}$$

one finds the momentum representation

$$\langle k'|\hat{V}_c|k \rangle = \frac{2}{\pi} \sum_{lm,l'm'} \int r^2 dr\; i^{l-l'} \left[j_{l'}(k'r) v(r) j_l(kr) \right]$$

$$\times \int d\Omega_r \left[Y_{l'm'}^*(\Omega_r) Y_{lm}(\Omega_r) \right] \left[Y_{l'm'}(\Omega_{k'}) Y_{lm}^*(\Omega_k) \right].$$

Integration over the solid angle using the orthogonality of the spherical harmonics and the definition

$$v_l(k',k) = \frac{2}{\pi} \int r^2 dr \left[j_l(k'r) v(r) j_l(kr) \right] \tag{3.43}$$

gives the result

$$\langle k'|\hat{V}_c|k \rangle = \sum_{lm} v_l(k',k) \left[Y_{lm}^*(\Omega_k) Y_{lm}(\Omega_{k'}) \right]. \tag{3.44}$$

An analogous calculation can be performed for the T-matrix elements

$$\langle k'|\hat{T}|k \rangle = \sum_{lm} T_l(k',k) Y_{lm}^*(\Omega_k) Y_{lm}(\Omega_{k'}). \tag{3.45}$$

In order to obtain an explicit expression for the partial T-matrix elements, one uses

$$\langle k'|\hat{T}|k \rangle = \langle k'|\hat{V}|\psi_k^{(+)} \rangle = \int d^3r\; \langle k'|r \rangle v(r) \langle r|\psi_k^{(+)} \rangle$$

and

$$\langle r|\psi_k^{(+)} \rangle = \left(\frac{2}{\pi}\right)^{1/2} \sum_{lm} (i)^l \frac{R_l(k,r)}{r} Y_{lm}(\Omega_r) Y_{lm}^*(\Omega_k).$$

---

[10] Note: $\frac{(2l+1)}{4\pi} P_l(\cos\theta) = \sum_m Y_{lm}^*(\Omega) Y_{lm}(\Omega') = \sum_m Y_{lm}(\Omega) Y_{lm}^*(\Omega')$

The result is

$$T_l(k', k) = \frac{2}{\pi} \int r^2 dr \; [j_l(k'r)v(r)R_l(k, r)]. \tag{3.46}$$

An integral equation for the partial T-matrix elements is obtained, if one inserts the expansions (3.44) and (3.45) into the Lippmann-Schwinger equation (2.28) for the T-matrix

$$\langle k'|\hat{T}|k\rangle = \langle k'|\hat{V}|k\rangle + \frac{2m_0}{\hbar^2} \int d^3k'' \frac{\langle k'|\hat{V}|k''\rangle\langle k''|\hat{T}|k\rangle}{(k_0^2 - k''^2 + i\epsilon)},$$

multiplies the resulting expression by

$$Y_{lm}^*(\Omega_{k'})Y_{lm}(\Omega_k) \qquad \text{with} \qquad l, m = \text{fixed}$$

and integrates over the solid angle in momentum space. The result is, for each angular momentum value $l = 0, 1, 2, \ldots$, a set of integral equations

$$T_l(k', k) = v_l(k', k) + \frac{2m_0}{\bar{h}^2} \int k''^2 dk'' \frac{v_l(k', k'')T_l(k'', k)}{(k_0^2 - k''^2 + i\epsilon)}. \tag{3.47}$$

With the partial wave expansion one can replace an integral equation in three dimensions by an infinite number of integral equations in one dimension. Of course, this is only useful, if one has to deal with only a few of the one-dimensional integral equations (e.g. in the case of short-range potentials).

For the on-shell T-matrix elements (1.19), (1.22) and (2.27) one finds with the addition theorem for the spherical harmonics

$$P_l(\cos\theta) = \frac{4\pi}{(2l+1)} \sum_m Y_{lm}^*(\Omega_k)Y_{lm}(\Omega_{k'}), \qquad \theta = \theta_{k,k'}$$

the representation in terms of the phase shifts $\delta_l(k)$

$$T_l(k', k)|_{\text{on}} \equiv T_l(k) = -\frac{\hbar^2}{\pi m_0 k} e^{i\delta_l(k)} \sin\delta_l(k). \tag{3.48}$$

Each wave number $k = k' = k_0$ is determined by the energy

$$E_0 = \frac{\hbar^2 k_0^2}{2m_0}.$$

The starting point for the discussion of the partial wave expansion of the K-matrix element $\langle k'|\hat{K}|k\rangle$ is the relation

$$\hat{K}|k\rangle = \hat{V}|\psi_k^{(s)}\rangle$$

and the Lippmann-Schwinger equation (3.41) for the K-matrix. A calculation following the pattern above yields

$$K_l(k', k) = v_l(k', k) + \frac{2m_0}{\hbar^2} \mathscr{P} \int k''^2 dk'' \, v_l(k', k'') \left\{ \frac{1}{k_0^2 - k''^2} \right\} K_l(k'', k).$$

(3.49)

For the discussion of this matrix element one still needs the partial wave expansion of the damping equation (3.39). For a computation of on-shell T-matrix elements, consideration of the equation

$$\langle k'|\hat{T}|k\rangle_{\mathrm{on}} = \langle k'|\hat{K}|k\rangle_{\mathrm{on}} - i\frac{\pi m_0 k}{\hbar^2} \int d\Omega_{k''} \langle k'|\hat{K}|k''\rangle_{\mathrm{on}} \langle k''|\hat{T}|k\rangle_{\mathrm{on}}$$

is sufficient. With $K_l(k', k)_{\mathrm{on}} = K_l(k)$ one finds after inserting the expansions

$$T_l(k) = K_l(k) - i\frac{\pi m_0 k}{\hbar^2} K_l(k) T_l(k).$$

(3.50)

The partial wave expansion of the damping equation leads to an algebraic equation instead of an integral equation. As the operator $\hat{K}$ is Hermitian, it follows that

$$K_l(k)^* = K_l(k).$$

The on-shell K-matrix elements are real. Resolution of (3.50) with respect to T_l

$$T_l(k) = \frac{K_l(k)}{1 + i\dfrac{\pi m_0 k}{\hbar^2} K_l(k)}$$

(3.51)

shows, that the on-shell T-matrix elements are complex (as required). Evaluation of the optical theorem (compare (2.42))

$$\hat{T}^\dagger - \hat{T} = 2\pi i \hat{T}^\dagger \delta(E_0 - \hat{H}_0)\hat{T}$$

in the partial wave representation leads to

$$\mathrm{Im}\, T_l(k) = -\frac{\pi m_0 k}{\hbar^2} |T_l(k)|^2.$$

(3.52)

It can be seen, that the optical theorem is always satisfied as a result of the relation (3.51), for example in the form

$$\mathrm{Im} T_l(k) = \mathrm{Im}\frac{K_l(k)}{1 + i\dfrac{\pi m_0 k}{\hbar^2}K_l(k)} = -\frac{\dfrac{\pi m_0 k}{\hbar^2}K_l(k)^2}{1 + \left(\dfrac{\pi m_0 k}{\hbar^2}\right)^2 K_l(k)^2}.$$

In addition, one obtains by resolving (3.50) with respect to $K_l(k)$

$$K_l(k) = \frac{T_l(k)}{1 - i\dfrac{\pi m_0 k}{\hbar^2}T_l(k)}$$

together with (3.48), a representation of $K_l(k)$ in terms of the phase shifts

$$K_l(k) = -\frac{\hbar^2}{\pi m_0 k}\frac{e^{i\delta_l}\sin\delta_l}{(1 + i\, e^{i\delta_l}\sin\delta_l)} = -\frac{\hbar^2}{\pi m_0 k}\tan\delta_l. \tag{3.53}$$

The *Heitler formalism* consists in solving the integral equation (3.49) approximately or iteratively in order to determine the on-shell K-matrix elements and then computing the on-shell T-matrix elements with (3.51) or the phase shifts with (3.53). Thus, for example, instead of the direct Born approximation

$$T_l(k)_{\mathrm{Born}} = v_l(k),$$

one obtains via the damping equation the result

$$T_l(k)_{\mathrm{Heitler}} = \frac{v_l(k)}{1 + i\dfrac{\pi m_0 k}{\hbar^2}v_l(k)},$$

which satisfies the optical theorem.

The partial wave expansion for the S-matrix elements is given directly in the form

$$\langle \boldsymbol{k}'|\hat{\mathrm{S}}|\boldsymbol{k}\rangle = \sum_{lm}\delta(k - k')\frac{S_l(k)}{k^2}Y_{lm}(\Omega_{k'})Y_{lm}^*(\Omega_k) \tag{3.54}$$

as only on-shell elements occur. The unitarity relation (3.36) then leads (details in Sect. 3.6.7) to

$$S_l^*(k)S_l(k) = 1. \tag{3.55}$$

From the relation (3.27) between the S-matrix elements and the T-matrix elements, one can extract the representation

$$S_l(k) = e^{2i\delta_l(k)}, \tag{3.56}$$

so that one finds with (1.22) and (1.24) for the cross section the (often used) result

$$\sigma = \frac{\pi}{k^2} \sum_l (2l + 1)|S_l(k) - 1|^2. \tag{3.57}$$

3.5.2   General Local Potentials: Selection Rules for the Partial Scattering Amplitudes

The discussion of scattering amplitudes in Sect. 1.4 for spin-dependent forces, as well as for spin-orbit or tensor forces, can be applied to the case of the three scattering matrices. The potentials, which normally play a role, are local in space and spin[11]

$$\langle r', 1'2'|\hat{V}|r, 12\rangle = \delta(r - r')\delta_{11'}\delta_{22'}\langle r, 12|\hat{V}|r, 12\rangle.$$

One distinguishes the following types of potentials (each with a spherically symmetric space part):

- Central potentials

$$\langle r, 12|\hat{V}_c|r, 12\rangle = v_c(r),$$

- Spin-spin potentials

$$\langle r, 12|\hat{V}_{ss}|r, 12\rangle = v_{ss}(r)\,(\hat{s}_1 \cdot \hat{s}_2),$$

- Spin-orbit potentials

$$\langle r, 12|\hat{V}_{sl}|r, 12\rangle = v_{sl}(r)\,\hat{l} \cdot (\hat{s}_1 + \hat{s}_2) = v_{sl}(r)\,\hat{l} \cdot \hat{S},$$

- Tensor potentials

$$\langle r, 12|\hat{V}_T|r, 12\rangle = v_T(r)\left[\frac{3(\hat{s}_1 \cdot r)(\hat{s}_2 \cdot r)}{r^2} - (\hat{s}_1 \cdot \hat{s}_2)\right].$$

---

[11] The spin degree of freedom of the particles $i = 1, 2$ is characterised by the operators s_i.

If one uses a wave function in coordinate space in the partial wave expansion multiplied by a spin part in the channel spin representation

$$\langle \boldsymbol{r}, 12 | \boldsymbol{k}, SM_S \rangle = \sum_{lm_l} \frac{R_l(k, r)}{r} Y_{lm_l}(\Omega_r) Y_{lm_l}^*(\Omega_k) \chi_{SM_S}(12),$$

one can write a general on-shell T-matrix element in the form

$$\langle \boldsymbol{k}', S'M_S' | \hat{T} | \boldsymbol{k}, SM_S \rangle |_{\text{on}} = \langle k\Omega_{k'}, S'M_S' | \hat{T} | k\Omega_k, SM_S \rangle$$

$$= \sum_{lm_l l'm_l'} \langle \Omega_{k'} | l'm_l' \rangle \langle kl'm_l', S'M_S' | \hat{T} | klm, SM_S \rangle \langle lm_l | \Omega_k \rangle$$

$$= \sum_{lml'm_l'} T_{l'm_l', S'M_S', lm_l \, SM_S}(k) \, Y_{l'm_l'}^*(\Omega_{k'}) Y_{lm_l}(\Omega_k).$$

$$(3.58)$$

Details are determined by the symmetry properties of the potentials, expressed in terms of associated commutators. In particular one finds

- for a central potential with

$$[\hat{V}_c, \hat{l}] = [\hat{V}_c, \hat{S}] = 0$$

the relation

$$[\hat{T}, \hat{l}] = [\hat{T}, \hat{S}] = 0$$

and from this (since the potential does not address the spin)

$$T_{l'm_l' \, S'M_S', lm_l \, SM_S}(k) = \delta_{ll'} \delta_{m_l m_l'} \delta_{SS'} \delta_{M_S M_S'} T_l(k).$$

- For a spin-spin potential one also has

$$[\hat{V}_{ss}, \hat{l}] = [\hat{V}_{ss}, \hat{S}] = 0$$

and thus

$$T_{l'm_l' \, S'M_S', lm_l \, SM_S}(k) = \delta_{ll'} \delta_{m_l m_l'} \delta_{SS'} \delta_{M_S M_S'} T_{lS}(k).$$

The partial T-matrix is different for the two spin channels, but there exist no transitions between the channels.

- For a spin-orbit potential one finds the following commutators

$$[\hat{V}_{sl}, \hat{l}^2] = [\hat{V}_{sl}, \hat{S}^2] = 0 \quad \text{but} \quad [V_{sl}, j] = 0 \quad \text{with} \quad \hat{j} = \hat{l} + \hat{S} \quad ,$$

as the commutators $[\hat{V}_{sl}, \hat{l}] = -[\hat{V}_{sl}, \hat{S}] \neq \mathbf{0}$ hold. For this reason one has to go from the

$$(l, m_l, S, M_S)-\text{basis to the } (j, m, l, S)-\text{basis}.$$

in order to obtain the matrix element

$$\langle k, j'm'l'S'|\hat{T}_{sl}|k, jmlS\rangle|_{\text{on}} = \delta_{jj'}\delta_{mm'}\delta_{ll'}\delta_{SS'}T_{jlS}(k).$$

Through the scattering by a spin-orbit potential, the spin and orbital angular momentum orientations can change. These changes are coupled as one has

$$m_l + M_S = m'_l + M'_S = m.$$

- A tensor potential can be characterised by

$$[V_T, S^2] = 0 \quad \text{and} \quad [V_T, j] = \mathbf{0}.$$

The operator $\hat{V}_T$ commutes with the operators for the components of the total angular momentum and the operator for the square of the magnitude of the channel spin operator. However, it does not commute with the square of the magnitude of the orbital angular momentum

$$[\hat{V}_T, \hat{l}^2] \neq 0.$$

This leads to the T-matrix element

$$\langle k, j'm'l'S'|\hat{T}_T|k, jmlS\rangle|_{\text{on}} = \delta_{jj'}\delta_{mm'}\delta_{SS'}T_{jS;ll'}(k).$$

The tensor potential mediates transitions between partial waves with different angular momentum quantum numbers l. The partial scattering matrix, as the scattering amplitude, have a more complicated structure.

The selection rules for scattering by spin-dependent potentials or with spin-dependent interactions will be addressed again in the next chapter (in Sect. 4.2) in the context of the discussion of conservation laws.

3.6   Detailed Calculations for Chap. 3

3.6.1   Existence of Asymptotic Time Limits

In a Hilbert space one can define the concepts of *strong* and *weak* convergence of sequences of vectors. A sequence of vectors $|\phi(t)\rangle$

- converges for $t \to t_0$ *strongly* if the length of the vectors tends to a fixed value

$$\lim_{t \to t_0} \langle \phi(t)|\phi(t)\rangle = A_s.$$

- The sequence converges for $t \to t_0$ *weakly* if the inner product of $|\phi(t)\rangle$ with any time-independent vector $|\chi\rangle$ tends towards a fixed value

$$\lim_{t \to t_0} \langle \chi|\phi(t)\rangle = A_\chi.$$

In order to prove the existence of the limits

$$|\psi_S(0)\rangle \equiv |\psi_I(0)\rangle = \lim_{t \to -\infty} \hat{U}_I^\dagger(t,0)|\psi_{in}\rangle = \lim_{t \to -\infty} \hat{U}_I(0,t)|\psi_{in}\rangle \tag{3.59}$$

for the incoming part and

$$|\psi_S(0)\rangle \equiv |\psi_I(0)\rangle = \lim_{t \to +\infty} \hat{U}_I^\dagger(t,0)|\psi_{out}\rangle$$

for the second half of the scattering process, one uses the integral equation (3.13)

$$\hat{U}_I(t,t_0) = \hat{1} - \frac{i}{\hbar} \int_{t_0}^t dt' \hat{V}_I(t') \hat{U}_I(t',t_0). \tag{3.60}$$

It is sufficient to present the arguments for one of the cases (e.g. the outgoing branch with $t_0 = 0$), since the detailed steps are equivalent in both cases.

3.6.2   Proof with a Wave Packet

The assumption is: The initial state is given in the form of a wave packet. Since the argument has to be carried through in ordinary space in this case, the Schrödinger picture is to be preferred. For an application of (3.60) one uses in the case of $t_0 = 0$

$$\hat{U}_I(t,0) = \hat{U}_0^\dagger(t,0)\hat{U}_S(t,0),$$
$$\hat{V}_I(t) = \hat{U}_0^\dagger(t,0)\hat{V}\hat{U}_0(t,0)$$

and finds the integral equation

$$\hat{U}_0^\dagger(t,0)\hat{U}_S(t,0) = \hat{1} - \frac{i}{\hbar}\int_0^t dt'\,\hat{U}_0^\dagger(t',0)\hat{V}\hat{U}_S(t',0).$$

Application of the adjoint equation to the initial state $|\psi_{in}\rangle$ yields

$$\hat{U}_S^\dagger(t,0)\hat{U}_0(t,0)|\psi_{in}\rangle = |\psi_{in}\rangle + \frac{i}{\hbar}\int_0^t dt'\,\hat{U}_S^\dagger(t',0)\hat{V}\hat{U}_0(t',0)|\psi_{in}\rangle. \qquad (3.61)$$

The limit $|\psi_S(0)\rangle$ exists, if the second term on the right-hand side converges for $t \to -\infty$ to a constant value. The condition for this is

$$\int_{-\infty}^0 dt\,||\hat{U}_S^\dagger(t,0)\hat{V}\hat{U}_0(t,0)|\psi_{in}\rangle|| = M < \infty$$

or in the sense of strong convergence

$$\int_{-\infty}^0 dt\left[\langle\psi_{in}|\hat{U}_0^\dagger(t,0)\hat{V}^2\hat{U}_0(t,0)|\psi_{in}\rangle\right]^{1/2} = M' < \infty.$$

The matrix element in the integrand has the form

$$\langle\psi_{in}|\hat{U}_0^\dagger(t,0)\hat{V}^2\hat{U}_0(t,0)|\psi_{in}\rangle = \int d^3r\,v(r)^2|\langle r|\hat{U}_0(t,0)|\psi_{in}\rangle|^2$$

for a local potential. For the estimate of the matrix element of the time development operator, one uses a wave packet, as for example a Gaussian packet[12]. For

$$\langle r|\psi_{in}\rangle = \exp\left[-\frac{(r-a)^2}{2s^2}\right]$$

one calculates

$$|\langle r|\hat{U}_0(t,0)|\psi_{in}\rangle|^2 = \left[1+\frac{\hbar^2 t^2}{m_0^2 s^4}\right]^{-3/2}\exp\left[-\frac{(r-a)^2}{s^2+\frac{\hbar^2 t^2}{m_0^2}}\right].$$

---

[12] There are only a few wave packets whose propagation can be calculated analytically. The estimate gives a similar result in each case.

This function can be majorised by the first factor, so that one obtains the estimate

$$\int_{-\infty}^{0} dt ||\hat{V}\hat{U}_0(t, 0)|\psi_{\text{in}}\rangle|| \leq \left[\int d^3r \, v(r)^2 \right] \left[\int_{-\infty}^{0} dt \left(1 + \frac{\hbar^2 t^2}{m_0^2 s^4} \right)^{-3/4} \right].$$

The time integral on the right-hand side has a finite value, due to the broadening of the wave packet. Convergence is therefore guaranteed if the first factor is convergent. This requires a potential, which is sufficiently continuous and increases for $r \to 0$ more slowly than $1/r^{(3/2-\delta)}$ and decreases for $r \to \infty$ faster than $1/r^{(3/2+\delta)}$.

3.6.3 Proof by Regularisation

Differentiation of the time development operator $\hat{U}_I(t_2, t_1)$ (use Eq. (3.10) with $t_0 = 0$)

$$\hat{U}_I(t_2, t_1) = \hat{U}_0^\dagger(t_2, 0)\hat{U}_S(t_2, t_1)\hat{U}_0(t_1, 0)$$

with respect to the variable t_1 yields

$$\frac{\partial \hat{U}_I(t_2, t_1)}{\partial t_1} = \frac{i}{\hbar}\hat{U}_0^\dagger(t_2, 0)\hat{U}_S(t_2, t_1)[\hat{H} - \hat{H}_0]\hat{U}_0(t_1, 0),$$

or

$$\frac{\partial \hat{U}_I(t_2, t_1)}{\partial t_1} = \frac{i}{\hbar}\hat{U}_0^\dagger(t_2, 0)\hat{U}_S(t_2, t_1)\hat{V}\hat{U}_0(t_1, 0)$$

$$= \frac{i}{\hbar}\hat{U}_0^\dagger(t_2, 0)\hat{U}_S(t_2, t_1)\hat{U}_0(t_1, 0)\hat{U}_0^\dagger(t_1, 0)\hat{V}\hat{U}_0(t_1, 0)$$

$$= \frac{i}{\hbar}\hat{U}_I(t_2, t_1)\hat{V}_I(t_1).$$

Integration leads to the integral equation

$$\hat{U}_I(t_2, t_1) = \hat{1} - \frac{i}{\hbar}\int_{t_1}^{t_2} dt' \, \hat{U}_I(t_2, t')\hat{V}_I(t').$$

This is an alternative to Eq. (3.13).

The state $|\psi_I(0)\rangle$ in the interaction picture at the time $t = 0$ can be generated from an initial state $|\mathbf{k}\rangle$ at the time $t = -\infty$ by

$$|\psi_I(0)\rangle = \hat{U}_I(0, -\infty)|\mathbf{k}\rangle.$$

The following argument shows that this state coincides with the state $|\psi_k^{(+)}\rangle$ of the stationary formulation, provided that the time development operator is used in the regularised form. Application of the integral equation for the time development operator and regularisation yields

$$|\psi_1(0)\rangle = |k\rangle - \lim_{\epsilon \to 0} \frac{i}{\hbar} \int_{-\infty}^{0} dt_2 \, \exp\{-\frac{\epsilon}{\hbar}|t_2|\}\hat{U}_1(0, t_2)\hat{V}_1(t_2)|k\rangle$$

$$= |k\rangle - \lim_{\epsilon \to 0} \frac{i}{\hbar} \int_{-\infty}^{0} dt_2 \, \exp\left\{\frac{i}{\hbar}(\hat{H} - i\epsilon)t_2\right\} \hat{V} \exp\left\{-\frac{i}{\hbar}\hat{H}_0 t_2\right\}|k\rangle$$

$$= |k\rangle - \lim_{\epsilon \to 0} \frac{i}{\hbar} \int_{-\infty}^{0} dt_2 \, \exp\left\{\frac{i}{\hbar}(\hat{H} - E(k) - i\epsilon)t_2\right\} \hat{V}|k\rangle.$$

The integration can now be performed. The result is

$$|\psi_1(0)\rangle = |k\rangle + \lim_{\epsilon \to 0} \frac{1}{(E(k) - \hat{H} + i\epsilon)}\hat{V}|k\rangle$$

$$\equiv |\psi_k^{(+)}\rangle.$$

By connecting the regularised time-dependent formulation to the stationary version, it can be shown that the limit (3.16) (similarly (3.17)) exists, if it is evaluated with sufficient care.

3.6.4 Perturbation Expansion of the S-Matrix: First and Second Order Contribution

The following tools are needed in this section:

- The Dirac identity (operator-valued)

$$\frac{1}{\hat{a} + i\epsilon} - \frac{1}{\hat{a} - i\epsilon} = -2\pi i \delta(\hat{a})$$

- and the definition of the δ-function in one dimension

$$\int_{-\infty}^{\infty} dx \, \exp(\pm ikx) = 2\pi \delta(k).$$

It can be shown by the (simple) evaluation of the contribution in *first order* that the operations $\epsilon \to 0$ and integration can be interchanged in this case. With

$$\hat{U}_0(t, 0) = \exp\left[-\frac{i}{\hbar}\hat{H}_0 t\right]$$

and

$$\hat{V}_I(t) = \exp\left[+ \frac{i}{\hbar}\hat{H}_0 t \right]\hat{V}\exp\left[- \frac{i}{\hbar}\hat{H}_0 t \right]$$

one obtains for the regularised S-matrix element

$$\langle k'|\hat{S}_\epsilon|k\rangle_{(1)} = -\frac{i}{\hbar}\int_{-\infty}^{\infty} dt_1 \exp\left[-\frac{\epsilon}{\hbar}|t_1| \right]$$

$$\exp\left[-\frac{i}{\hbar}(E(k) - E(k'))t_1 \right]\langle k'|\hat{V}|k\rangle.$$

For the explicit evaluation, the time integral must be divided into two parts as a result of the regularisation

$$I = \int_{-\infty}^{0} dt_1 \exp\left[-\frac{i}{\hbar}(E(k) - E(k') + i\epsilon)t_1 \right]$$

$$+ \int_{0}^{\infty} dt_1 \exp\left[-\frac{i}{\hbar}(E(k) - E(k') - i\epsilon)t_1 \right].$$

The result of the standard integration

$$I = -\frac{\hbar}{i}\left\{ \frac{1}{(E(k) - E(k') + i\epsilon)} - \frac{1}{(E(k) - E(k') - i\epsilon)} \right\}$$

yields in the limiting case $\epsilon \to 0$ with the Dirac identity the S-matrix element

$$\langle k'|\hat{S}|k\rangle_{(1)} = -2\pi i\, \delta(E(k) - E(k'))\langle k'|\,\hat{V}|k\rangle.$$

If the operations are interchanged, the same result follows after substitution due to

$$\langle k'|\hat{S}|k\rangle_{(1)} = -i\int_{-\infty}^{\infty} dt \exp\left[-i(E(k) - E(k'))t \right]\langle k'|\,\hat{V}|k\rangle.$$

The evaluation of the contribution in *second order* is somewhat more tedious. One should calculate

$$\langle k'|\hat{S}|k\rangle_{(2)} = \left(-\frac{i}{\hbar}\right)^2 \lim_{\epsilon \to 0}\left\{ \int d^3k'' \int_{-\infty}^{\infty} dt_2 \ \langle k'|e^{\frac{i}{\hbar}\hat{H}_0 t_2}\hat{V}e^{-\frac{i}{\hbar}\hat{H}_0 t_2}|k''\rangle \right.$$

$$\left. \times e^{-\frac{\epsilon}{\hbar}|t_2|}\int_{-\infty}^{t_2} dt_1 \ e^{-\frac{\epsilon}{\hbar}|t_1|}\ \langle k''|e^{\frac{i}{\hbar}\hat{H}_0 t_1}\hat{V}e^{-\frac{i}{\hbar}\hat{H}_0 t_1}|k\rangle \right\},$$

or after application of the time development operators on the plane wave states

$$\langle k'|\hat{S}|k\rangle_{(2)} = \left(-\frac{i}{\hbar}\right)^2 \lim_{\epsilon \to 0} \left\{ \int d^3k'' \langle k'|\hat{V}|k''\rangle\langle k''|\hat{V}|k\rangle \right.$$

$$\times \int_{-\infty}^{\infty} dt_2\ e^{\frac{i}{\hbar}(E(k')-E(k''))t_2}\ e^{-\frac{\epsilon}{\hbar}|t_2|} \qquad (3.62)$$

$$\left. \times \int_{-\infty}^{t_2} dt_1\ e^{-\frac{\epsilon}{\hbar}|t_1|}e^{\frac{i}{\hbar}(E(k'')-E(k))t_1} \right\}.$$

For the evaluation of the double integral in time

$$I = \int_{-\infty}^{\infty} dt_2\ e^{\frac{i}{\hbar}(E(k')-E(k''))t_2}\ e^{-\frac{\epsilon}{\hbar}|t_2|} \int_{-\infty}^{t_2} dt_1\ e^{-\frac{\epsilon}{\hbar}|t_1|}e^{\frac{i}{\hbar}(E(k'')-E(k))t_1}$$

$$\equiv \int_{-\infty}^{\infty} dt_2\ f_2(t_2) \int_{-\infty}^{t_2} dt_1\ f_1(t_1)$$

one has to divide the area of integration due to the structure of the convergence factors into three subareas with the limits (Fig. 3.1)

- $-\infty \le t_1 \le t_2 \le 0,$
- $-\infty \le t_1 \le 0$ and $0 \le t_2 \le \infty,$
- $0 \le t_1 \le t_2 \le \infty.$

The integral I is thus given by

$$I = I_1 + I_2 + I_3$$

Fig. 3.1 Subdivision of the integration domain for the computation of $\langle k'|\hat{S}|k\rangle_{(2)}$

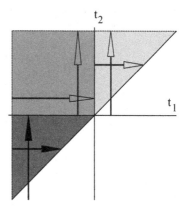

with

$$I_1 = \int_{-\infty}^{0} dt_2 \, f_2(t_2) \int_{-\infty}^{t_2} dt_1 \, f_1(t_1),$$

$$I_2 = \int_{0}^{\infty} dt_2 \, f_2(t_2) \int_{-\infty}^{0} dt_1 \, f_1(t_1),$$

$$I_3 = \int_{0}^{\infty} dt_2 \, f_2(t_2) \int_{0}^{t_2} dt_1 \, f_1(t_1).$$

The individual integrals are: The inner integral in I_1

$$I_{11}(t_2) = \int_{-\infty}^{t_2} dt_1 \, e^{\frac{i}{\hbar}(E(k'')-E(k)-i\epsilon)t_1} = \left(\frac{\hbar}{i}\right) \frac{e^{\frac{i}{\hbar}(E(k'')-E(k)-i\epsilon)t_2}}{(E(k'') - E(k) - i\epsilon)}$$

leads to I_1 with

$$I_1 = \left(\frac{\hbar}{i}\right) \int_{-\infty}^{0} dt_2 \, \frac{e^{\frac{i}{\hbar}(E(k')-E(k)-2i\epsilon)t_2}}{(E(k'') - E(k) - i\epsilon)}$$

$$= \left(\frac{\hbar}{i}\right)^2 \frac{1}{(E(k'') - E(k) - i\epsilon)(E(k') - E(k) - 2i\epsilon)}.$$

The integral I_2 factorises

$$I_2 = \int_{0}^{\infty} dt_2 \, e^{\frac{i}{\hbar}(E(k')-E(k'')+i\epsilon)t_2} \int_{-\infty}^{0} dt_1 \, e^{\frac{i}{\hbar}(E(k'')-E(k)-i\epsilon)t_1}$$

$$= -\left(\frac{\hbar}{i}\right)^2 \frac{1}{(E(k') - E(k'') + i\epsilon)(E(k'') - E(k) - i\epsilon)}.$$

The internal integration in the integral I_3 extends over a finite range. No convergence factor is required. Alternatively such a factor can be chosen arbitrarily as the factor ϵ goes to zero $\epsilon \to 0$. The following choice proves to be convenient

$$I_{31} = \int_{0}^{t_2} dt_1 \, e^{\frac{i}{\hbar}(E(k'')-E(k)-i\epsilon)t_1} = \left(\frac{\hbar}{i}\right) \frac{1}{(E(k'') - E(k) - i\epsilon)}$$

$$\times \left\{ e^{\frac{i}{\hbar}(E(k'')-E(k)-i\epsilon)t_2} - 1 \right\}.$$

The outer integration with different convergence factors for the two terms gives

$$
I_3 = \left(\frac{\hbar}{i}\right) \frac{1}{(E(k'') - E(k) - i\epsilon)} \int_0^\infty dt_2 \left\{ e^{\frac{i}{\hbar}(E(k') - E(k) + 2i\epsilon)t_2} \right.
$$
$$
\left. - e^{\frac{i}{\hbar}(E(k') - E(k'') + i\epsilon)t_2} \right\}
$$
$$
= \left(\frac{\hbar}{i}\right)^2 \frac{1}{(E(k'') - E(k) - i\epsilon)} \left\{ -\frac{1}{(E(k') - E(k) + 2i\epsilon)} \right.
$$
$$
\left. + \frac{1}{(E(k') - E(k'') + i\epsilon)} \right\}.
$$

The result for the integral I is therefore

$$
\left(\frac{i}{\hbar}\right)^2 I = -\frac{1}{(E(k) - E(k'') + i\epsilon)(E(k') - E(k) - 2i\epsilon)}
$$
$$
+ \frac{1}{(E(k') - E(k'') + i\epsilon)(E(k) - E(k'') + i\epsilon)}
$$
$$
+ \frac{1}{(E(k) - E(k'') + i\epsilon)(E(k') - E(k) + 2i\epsilon)}
$$
$$
- \frac{1}{(E(k') - E(k'') + i\epsilon)(E(k) - E(k'') + i\epsilon)}
$$
$$
= \frac{1}{(E(k) - E(k'') + i\epsilon)} \left\{ \frac{1}{(E(k') - E(k) + 2i\epsilon)} - \frac{1}{(E(k') - E(k) - 2i\epsilon)} \right\}
$$
$$
= -2\pi i \frac{\delta(E(k) - E(k'))}{(E(k) - E(k'') + i\epsilon)}.
$$

In the last step the Dirac identity has been used. Inserting this result in (3.62), one obtains, in agreement with the result in Sect. 3.3.2

$$
\langle k'|\hat{S}|k\rangle_{(2)} = -2\pi i\delta(E(k) - E(k')) \left\langle k' \left| \hat{V} \frac{1}{(E(k) - \hat{H}_0 + i\epsilon)} \hat{V} \right| k \right\rangle.
$$

3.6.5   Relation Between the S- and the T-Matrix, General Argument

The argumentation is based on the remarks in Sect. 3.2.2. The S-matrix element

$$
\langle k'|\hat{S}|k\rangle = \langle k'|\hat{U}_I(\infty, -\infty)|k\rangle
$$

must be evaluated with the regularised form of the integral equation (3.13) for the time development operator, which yields

$$\langle k'|\hat{S}|k\rangle = \delta(k-k') - \lim_{\epsilon \to 0} \frac{i}{\hbar} \int_{-\infty}^{\infty} dt_2 \, e^{-\frac{\epsilon}{\hbar}|t_2|} \langle k'|\hat{V}_I(t_2)\hat{U}_I(t_2,-\infty)|k\rangle.$$

One now uses the multiplication theorem

$$\hat{U}_I(t_2,-\infty) = \hat{U}_I(t_2,0)\hat{U}_I(0,-\infty),$$

and (see Sect. 3.6.3)

$$\hat{U}_I(0,-\infty)|k\rangle = |\psi_k^{(+)}\rangle$$

as well as

$$\begin{aligned}
\hat{U}_I(t_2,0)|\psi_k^{(+)}\rangle &= \hat{U}_0^\dagger(t_2,0)\hat{U}_S(t_2,0)|\psi_k^{(+)}\rangle \\
&= e^{\frac{i}{\hbar}\hat{H}_0 t_2} e^{-\frac{i}{\hbar}\hat{H}t_2}|\psi_k^{(+)}\rangle = e^{\frac{i}{\hbar}\hat{H}_0 t_2} e^{-\frac{i}{\hbar}E(k)t_2}|\psi_k^{(+)}\rangle.
\end{aligned}$$

With these steps one finds

$$\langle k'|\hat{S}|k\rangle = \delta(k-k') - \lim_{\epsilon \to 0} \frac{i}{\hbar} \int_{-\infty}^{\infty} dt_2 \, e^{-\frac{\epsilon}{\hbar}|t_2|} e^{-\frac{i}{\hbar}(E(k)-E(k'))t_2} \langle k'|\hat{V}|\psi_k^{(+)}\rangle.$$

The limiting case $\epsilon \to 0$ is not critical, so that one can use one of the definitions of the δ-function and obtain

$$\begin{aligned}
\langle k'|\hat{S}|k\rangle &= \delta(k-k') - 2\pi i\delta(E(k)-E(k'))\langle k'|\hat{V}|\psi_k^{(+)}\rangle \\
&= \delta(k-k') - 2\pi i\delta(E(k)-E(k'))\langle k'|\hat{T}|k\rangle.
\end{aligned}$$

3.6.6 Orthogonality of Scattering States

For an explicit discussion of the orthogonality relations

$$\langle \psi_k^{(+)}|\psi_{k'}^{(\pm)}\rangle$$

one first uses the Lippmann-Schwinger equation (2.22) for $|\psi_k^{(+)}\rangle$ and obtains

$$\begin{aligned}
\langle \psi_k^{(+)}|\psi_{k'}^{(\pm)}\rangle &= \langle k|\psi_{k'}^{(\pm)}\rangle + \left\langle k \left| \hat{V} \frac{1}{(E(k)-\hat{H}-i\epsilon)} \right| \psi_{k'}^{(\pm)} \right\rangle \\
&= \langle k|\psi_{k'}^{(\pm)}\rangle + \frac{1}{(E(k)-E(k')-i\epsilon)} \langle k|\hat{V}|\psi_{k'}^{(\pm)}\rangle.
\end{aligned}$$

The last step follows because of

$$\hat{H}|\psi_{k'}^{(\pm)}\rangle = E(k')|\psi_{k'}^{(\pm)}\rangle.$$

Use now the Lippmann-Schwinger equation (2.13) for $|\psi_{k'}^{(\pm)}\rangle$ in the first term

$$\langle \psi_k^{(+)}|\psi_{k'}^{(\pm)}\rangle = \langle k|k'\rangle + \left\langle k \left| \frac{1}{(E(k') - \hat{H}_0 \pm i\epsilon)} \hat{V} \right| \psi_{k'}^{(\pm)} \right\rangle$$

$$+ \frac{1}{(E(k) - E(k') - i\epsilon)} \langle k|\hat{V}|\psi_{k'}^{(\pm)}\rangle$$

$$= \langle k|k'\rangle + \left[\frac{1}{(E(k') - E(k) \pm i\epsilon)} + \frac{1}{(E(k) - E(k') - i\epsilon)} \right]$$

$$\times \langle k|\hat{V}|\psi_{k'}^{(\pm)}\rangle.$$

For the expression in the square brackets one finds

$$[\ldots] = \left[\frac{1}{(E(k) - E(k') - i\epsilon)} - \frac{1}{(E(k) - E(k') - i\epsilon)} \right] = 0$$

or

$$= \left[\frac{1}{(E(k) - E(k') - i\epsilon)} - \frac{1}{(E(k) - E(k') + i\epsilon)} \right] \neq 0.$$

It vanishes for equal boundary conditions but not for unequal boundary conditions. Orthogonality can only be guaranteed for equal boundary conditions.

3.6.7 Unitarity Relation for the Partial S-Matrix Elements

The starting point is the general unitarity relation (3.36)

$$\int d^3k'' \langle k'|\hat{S}^\dagger|k''\rangle \langle k''|\hat{S}|k\rangle = \delta(k - k').$$

Insert the expansion (3.54) and obtain for the left-hand side

$$LHS = \int (k'')^2 dk'' \int d\Omega_{k''} \sum_{lm} \sum_{l'm'} \delta(k'' - k') \frac{S_l(k'')^*}{(k'')^2} Y_{lm}^*(\Omega_{k''})$$

$$\times Y_{lm}(\Omega_{k'}) \delta(k'' - k) \frac{S_{l'}(k)}{k^2} Y_{l'm'}(\Omega_{k''}) Y_{l'm'}^*(\Omega_k).$$

Integration over the double-primed coordinates gives

$$LHS = \frac{\delta(k - k')}{k^2} \sum_l S_l(k)^* S_l(k) \sum_m Y_{lm}^*(\Omega_k) Y_{lm}(\Omega_{k'}).$$

The three-dimensional delta function on the right-hand side can be written in the form

$$\delta(\boldsymbol{k} - \boldsymbol{k}') = \frac{\delta(k - k')}{k^2} \sum_{lm} Y_{lm}^*(\Omega_k) Y_{lm}(\Omega_{k'}).$$

Comparison (or multiplication with suitable spherical harmonics and integration over the two solid angles) gives

$$S_l(k)^* S_l(k) = 1.$$

Literature in Chap. 3

1. C. Møller, Danske Videnskab. Selskab, Mat-fys. Medd. **23**, p. 1 (1948)
2. J. A. Wheeler, Phys. Rev. **52**, p. 1107 (1937)
3. W. Heisenberg, Z. Phys. **120**, p. 513 (1943)

Conservation Laws in Scattering Theory

4

Conservation laws were already addressed in Sect. 1.4.2 with a discussion of selection rules for the elastic scattering of two particles as a consequence of the commutation relations of the Hamiltonian with angular momentum operators. The Hamiltonian

$$\hat{H} = \hat{T} + \hat{V}$$

consisted of the kinetic energy for the relative motion of the particles and a spin-dependent two-particle interaction

$$\langle r', \sigma_1', \sigma_2' | \hat{V} | r, \sigma_1, \sigma_2 \rangle = \delta(r' - r)\delta_{\sigma_1',\sigma_1}\delta_{\sigma_2',\sigma_2} v(r, \sigma_1, \sigma_2).$$

In this chapter a more general Hamiltonian will be considered, in which the operator $\hat{H}_0$ contains additional one-particle potential operators $\hat{U}$ besides the kinetic energy

$$\hat{H} = \hat{H}_0 + \hat{V}$$
$$\hat{H}_0 = \hat{T} + \hat{U}.$$

The symmetry operations in ordinary and spin space (or in momentum and spin space) of quantum collision systems include, besides space rotations, also displacements of the system, space reflections and time reversal—i.e. all symmetry operations of interest in the nonrelativistic domain. The results obtained depend only on the symmetry properties of the Hamiltonian and are independent of its special form. They apply as well to inelastic processes, e.g. to processes in which excited particles or more than two particles appear in the exit channels.

The effect of the symmetry operations are determined by the action of operators on the quantum states $|\psi\rangle$ of the system. These operators are determined by a

© Springer-Verlag GmbH Germany, part of Springer Nature 2022
R. M. Dreizler et al., *Quantum Collision Theory of Nonrelativistic Particles*,
https://doi.org/10.1007/978-3-662-65591-7_4

parameter or a set of parameters a which characterise the operation

$$|\psi_a\rangle = \hat{O}(a)|\psi\rangle.$$

If an operator $\hat{O}(a)$, for which the validity of $\hat{O}^\dagger(a)\hat{O}(a) = 1$ is assumed, commutes with the Hamiltonian of the system

$$\hat{O}(a)\hat{H} = \hat{H}\hat{O}(a) \quad \text{or} \quad \hat{O}^\dagger(a)\hat{H}\hat{O}(a) = \hat{H},$$

the result of experiments can be determined by both the stationary state $|\psi\rangle$ as well as by the stationary state $|\psi_a\rangle$. Physically relevant quantities, such as cross sections, are not changed by symmetry operations, a fact, which reflects the validity of conservation laws.

A pattern for the proof of the validity of conservation laws in scattering experiments involves, for each of the four symmetry operations considered here, the following steps:

1. Investigate the commutation relations of the operators $\hat{O}(a)$ with the Hamiltonian $\hat{H}$ and the Møller operators $\hat{\Omega}_\pm$, which are defined in (3.18).
2. If the Møller operators commute with the operator $\hat{O}(a)$, then this operator also commutes with the S-matrix operator $\hat{S}$, given in (3.22).
3. The S-matrix elements with respect to the original states and the transformed states are equal in this case

$$\langle\psi_a'|\hat{S}|\psi_a\rangle = \langle\psi'|\hat{O}^\dagger(a)\hat{S}\hat{O}(a)|\psi\rangle = \langle\psi'|\hat{S}|\psi\rangle.$$

4. It then follows, on the basis of the relation (compare (2.27) and (3.27))

$$\langle k'|(\hat{S}-1)|k\rangle = \frac{i\hbar^2}{2\pi m_0}\delta(E(k) - E(k'))f(k \to k'),$$

that the scattering amplitudes $f(k \to k')$ are not changed by the transformation of the states.

This chapter begins with a brief classification of the operations, which are addressed. A complete list of their properties as well as explicit examples can be found in textbooks of quantum mechanics and functional analysis. Subsequent to the classification, the invariance of the scattering amplitudes—or more generally of the S-matrix elements—is established.

4.1   The Classification of Operators

In quantum mechanics an operator is called *linear*, if it transforms a linear combination of elements of a Hilbert space $\sum_{i=1}^{N} a_i |\psi_i\rangle$ into a linear combination of the transformed elements. A linear operator $\hat{O}_{\text{lin}}$ thus satisfies the relation

$$\hat{O}_{\text{lin}} \left(\sum_{i=1}^{N} a_i |\psi_i\rangle \right) = \sum_{i=1}^{N} a_i \left(\hat{O}_{\text{lin}} |\psi_i\rangle \right) \tag{4.1}$$

for arbitrary complex numbers a_i and for arbitrary elements $|\psi_i\rangle$ of the Hilbert space.

An operator $\hat{O}_{\text{alin}}$ is referred to as *antilinear*, if it satisfies the relation

$$\hat{O}_{\text{alin}} \left(\sum_{i=1}^{N} a_i |\psi_i\rangle \right) = \sum_{i=1}^{N} a_i^* \left(\hat{O}_{\text{alin}} |\psi_i\rangle \right). \tag{4.2}$$

Instead of the coefficients a_i the complex conjugate quantities a_i^* appear in the transformed linear combination.

A class of linear operators of particular interest in physics are the *Hermitian* operators. The definition of these operators, which is used here, is: a linear operator $\hat{O}_{\text{herm}}$ is called Hermitian, if it satisfies the additional relation

$$\langle (\hat{O}_{\text{herm}} \psi_1) | \psi_2 \rangle = \langle \psi_1 | (\hat{O}_{\text{herm}} \psi_2) \rangle. \tag{4.3}$$

An explicit example for a one-particle situation in ordinary space is

$$\int d^3 r \left[O_{\text{herm}}(\boldsymbol{r}, \nabla) \psi_1(\boldsymbol{r}) \right]^* \psi_2(\boldsymbol{r}) = \int d^3 r \, \psi_1(\boldsymbol{r})^* \left[O_{\text{herm}}(\boldsymbol{r}, \nabla) \psi_2(\boldsymbol{r}) \right].$$

This property guarantees, that expectation values of Hermitian operators are real and can be identified with measured values of physical quantities.[1] The importance of Hermitian operators is emphasised by the fact, that they are useful for the formulation of eigenvalue problems of quantum mechanics, as they preserve properties such as the orthonormality and the completeness of solutions.

An operator $\hat{O}_{\text{herm}}^{\dagger}$, which is defined by the relation

$$\int d^3 r \left[O_{\text{herm}}(\boldsymbol{r}, \nabla) \psi_1(\boldsymbol{r}) \right]^* \psi_2(\boldsymbol{r}) = \int d^3 r \, \psi_1^*(\boldsymbol{r}) \left[O_{\text{herm}}^{\dagger}(\boldsymbol{r}, \nabla) \psi_2(\boldsymbol{r}) \right] \tag{4.4}$$

---

[1] Alternative definitions are found in the mathematical literature.

or formally by

$$\langle (\hat{O}_{herm}\psi_1)|\psi_2 \rangle = \langle \psi_1 | \hat{O}^{\dagger}_{herm} | \psi_2 \rangle,$$

is called the *Hermitian adjoint* operator to the operator $\hat{O}_{herm}$. An operator for which the relation

$$\hat{O}^{\dagger}_{herm} \equiv \hat{O}_{herm} \tag{4.5}$$

is satisfied, is called *self-adjoint*.

Linear operators, for which the inverse operator is equal to the adjoint operator, are called *unitary*. The condition

$$\hat{O}^{-1}_{uni} = \hat{O}^{\dagger}_{uni} \tag{4.6}$$

can also be written in the form

$$\hat{O}_{uni}\hat{O}^{\dagger}_{uni} = \hat{O}^{\dagger}_{uni}\hat{O}_{uni} = \hat{1}.$$

In this case one finds

$$\langle \hat{O}_{uni}\psi_1 | \hat{O}_{uni}\psi_2 \rangle = \langle \psi_1 | \hat{O}^{\dagger}_{uni}\hat{O}_{uni} | \psi_2 \rangle = \langle \psi_1 | \psi_2 \rangle.$$

If a unitary operator is applied to all elements of a vector space, the *lengths of the vectors* and the *angles between the vectors* are not changed, as they are expressed (for example) by bra-ket inner products. If one uses the denumerable (or denumerably infinite) eigenvectors of a Hermitian eigenvalue problem

$$|\psi_1\rangle, |\psi_2\rangle, |\psi_3\rangle, \ldots,$$

one can generate a matrix representation of physical quantities, which are characterised by an operator $\hat{G}$ with the matrix elements

$$G_{m,n} = \langle \psi_m | \hat{G} | \psi_n \rangle.$$

Such representations can be used to treat quantum mechanical problems by algebraic methods.[2]

---

[2] In the mathematical literature, there is usually the additional requirement that unitary operators be bounded, i.e. that the relation $\left| \langle \psi | \hat{O}_{uni} | \psi \rangle \right| < M$ holds.

For the discussion of time reversal *anti-unitary* operators are needed. An anti-unitary operator $\hat{O}_{\text{auni}}$ is antilinear and is formally defined by

$$\langle \hat{O}_{\text{auni}} \psi_1 | \hat{O}_{\text{auni}} \psi_2 \rangle = \langle \psi_1 | \psi_2 \rangle^*. \tag{4.7}$$

The explicit form for one-particle systems is

$$\int \mathrm{d}^3 r \big[\hat{O}^{\dagger}_{\text{auni}} \psi_1^*(r) \big] \big[\hat{O}_{\text{auni}} \psi_2(r) \big] = \Big[\int \mathrm{d}^3 r\, \psi_1^*(r) \psi_2(r) \Big]^*.$$

An alternative formal characterisation is (as defined in (4.2))

$$\hat{O}_{\text{auni}} \big(a | \psi \rangle \big) = a^* \big(\hat{O}_{\text{auni}} | \psi \rangle \big).$$

The operator $\hat{O}^{\dagger}_{\text{auni}}$, adjoint to $\hat{O}_{\text{auni}}$, is defined by

$$\langle \hat{O}_{\text{auni}} \psi_1 | \psi_2 \rangle = \langle \psi_1 | \hat{O}^{\dagger}_{\text{auni}} | \psi_2 \rangle^*.$$

If the two states are equal ($|\psi_2\rangle = |\psi_1\rangle$) and normalised, the definition (4.7) leads to

$$\langle \psi_1 | \hat{O}^{\dagger}_{\text{auni}} \hat{O}_{\text{auni}} | \psi_1 \rangle = \langle \psi_1 | \psi_1 \rangle^* = 1.$$

As $|\psi_1\rangle$ can be chosen arbitrarily, the operator relation

$$\hat{O}^{\dagger}_{\text{auni}} \hat{O}_{\text{auni}} = \hat{1} \tag{4.8}$$

follows.

The simplest situation, in which the classes of operators characterised above, are discussed in scattering theory, is the treatment of elastic scattering of spinless particles.

4.2   Elastic Scattering: Spinless Particles

4.2.1   Translation: Conservation of Momentum

A translation of the scattering arrangement by the distance a is described by a unitary displacement operator in three-dimensional space $\hat{O}_{\text{trans}}(a)$

$$\hat{O} \longrightarrow \hat{O}_{\text{trans}}(a) = \mathrm{e}^{-\mathrm{i} a \cdot \hat{P}/\hbar}.$$

The operator $\hat{P}$ represents the operator for the total momentum of the two-particle system. The representation of the operator in ordinary space is

$$\langle r'_1, r'_2 | \hat{O}_{\text{trans}}(a) | r_1, r_2 \rangle = \delta(r_1 - r'_1)\delta(r_2 - r'_2)e^{-a\cdot(\nabla_1 + \nabla_2)}.$$

The transformed wave function

$$\langle r_1, r_2 | \hat{O}_{\text{trans}}(a) | \psi \rangle = e^{-a\cdot(\nabla_1 + \nabla_2)}\langle r_1, r_2 | \psi \rangle \tag{4.9}$$

corresponds to the Taylor expansion of $\psi(r_1 - a, r_2 - a)$ (details are given in Sect. 4.4.1).

The momentum operator $\hat{P}$ commutes with both $\hat{H}$ as well as with $\hat{H}_0$. In consequence, it also commutes with the Møller operators $\hat{\Omega}_{\pm}$ and with the S-matrix operator $\hat{S}$. The S-matrix element with two scattering states has the form

$$\langle \psi'(a) | \hat{S} | \psi(a) \rangle = \langle \psi' | \hat{O}^{\dagger}_{\text{trans}}(a)\hat{S}\hat{O}_{\text{trans}}(a) | \psi \rangle = \langle \psi' | \hat{S} | \psi \rangle.$$

This equation describes the fact, that an experiment gives the same result at any point of space.

If one characterises the motion of two particles by plane waves with

$$\hat{P} | k_1, k_2 \rangle = \hbar(k_1 + k_2) | k_1, k_2 \rangle,$$

a matrix element of the commutator of $\hat{P}$ and $\hat{S}$

$$\langle k'_1, k'_2 | [\hat{P}, \hat{S}] | k_1, k_2 \rangle = 0$$

can be calculated by separation into relative and centre of mass motion. On the basis of the transformations

$$K = k_1 + k_2, \qquad k_{\text{rel}} = \frac{m_{20}k_1 - m_{10}k_2}{m_{10} + m_{20}}$$

and

$$R = \frac{m_{10}r_1 + m_{20}r_2}{m_{10} + m_{20}}, \qquad r = r_1 - r_2$$

one finds for the arguments of the wave functions

$$k_1 \cdot r_1 + k_2 \cdot r_2 = k_{\text{rel}} \cdot r + K \cdot R.$$

The operator for the total momentum acts only on the part describing the motion of the centre of mass, so that the statement

$$\langle k'_{\mathrm{rel}}, K' | [\hat{P}, \hat{S}] | k_{\mathrm{rel}}, K \rangle = K \, \delta(K' - K) \langle k'_{\mathrm{rel}}, K | \hat{S} | k_{\mathrm{rel}}, K \rangle = 0$$

follows. The total momentum is a conserved quantity. In a centre of mass system one has $K = 0$. The S-matrix element in this system is determined only by the relative motion with $|k_{\mathrm{rel}}| = |k'_{\mathrm{rel}}|$. The S-matrix element is, as expected, on-shell.

In a many-particle system with $K = \sum_i k_i$ one evaluates the matrix element of the commutator $[\hat{P}, \hat{S}]$ in the same way

$$\langle k'_1, k'_2, \ldots | [\hat{P}, \hat{S}] | k_1, k_2, \ldots \rangle = K \, \delta(K - K')$$

$$\times \langle \tilde{k}'_1, \tilde{k}'_2, \ldots, K | \hat{S} | \tilde{k}_1, \tilde{k}_2, \ldots, K \rangle = 0,$$

so that conservation of momentum is also guaranteed. The wave numbers $\tilde{k}$ are referred to the *centre of mass* system.

4.2.2 Rotation: Conservation of Angular Momentum

A unitary operator, which describes arbitrary spatial rotations can be represented by the three Euler angles[3] $\Omega \rightarrow \{\alpha, \beta, \gamma\}$

$$\hat{O}_{\mathrm{rot}}(\Omega) = e^{-i\alpha \hat{J}_z/\hbar} e^{-i\beta \hat{J}_y/\hbar} e^{-i\gamma \hat{J}_z/\hbar}.$$

A simpler form is obtained by specifying an axis of rotation, which is represented by a unit vector n, and a rotation through an angle ω

$$\hat{O}_{\mathrm{rot}}(n, \omega) = e^{-i\omega(n \cdot \hat{J})/\hbar}.$$

Alternatively, a vectorial angle of rotation $\boldsymbol{\omega} = \omega \, \boldsymbol{n}$ can be used. The total angular momentum $\mathbf{J}$ of two particles, each with the spin s_i, is the sum of the orbital angular momentum of the centre of mass, the orbital angular momentum of the relative motion and the spin of the two particles

$$\hat{J} = \hat{R} \times \hat{P} + \hat{r} \times \hat{p} + \hat{s}_1 + \hat{s}_2.$$

---

[3] See Dreizler/Lüdde, Vol. 1, Sect. 6.3.5 and M. E. Rose: Elementary Theory of Angular Momentum. J. Wiley, New York (1957); reprinted by Dover Publications, New York (1995), p. 51.

In elastic scattering with a spin-independent interaction[4] only the relative motion plays a role, so that one can restrict the discussion to

$$\hat{\boldsymbol{j}}_{\text{rel}} = \hat{\boldsymbol{l}}_{\text{rel}} = \hat{\boldsymbol{r}} \times \hat{\boldsymbol{p}}. \tag{4.10}$$

If the interaction potential is spherically symmetric $v(\boldsymbol{r}) = v(r)$, the scattering operator $\hat{\mathsf{S}}$ commutes with $\hat{O}_{\text{rot}}(\omega) = \text{e}^{-i\omega(\boldsymbol{n}\cdot\hat{\boldsymbol{l}}_{\text{rel}})/\hbar}$

$$\hat{O}_{\text{rot}}(\omega)\hat{\mathsf{S}} = \hat{\mathsf{S}}\hat{O}_{\text{rot}}(\omega).$$

A consequence of this commutation relation is the statement

$$\langle k_1|\hat{\mathsf{S}}|k_2\rangle = \langle k_{R1}|\hat{\mathsf{S}}|k_{R2}\rangle \quad \text{with} \quad |k_R\rangle = \hat{O}_{\text{rot}}|k\rangle. \tag{4.11}$$

The S-matrix element of two momentum states is equal to the matrix element between the corresponding momentum states in a system that was rotated by an angle ω about the axis $\boldsymbol{n}$. In elastic scattering, the magnitudes of the momenta of the states in (4.11) are pairwise equal. As the angle between the momentum vectors on the left- and right-hand sides of Eq. (4.11) does not change with a rotation, it can be concluded that the associated scattering amplitude for the scattering by a spin-independent potential with spherical symmetry depends only on two quantities, namely the energy and a scattering angle θ. This angle can be measured, for example, with respect to the z-axis, the direction of the incident beam

$$f(k \to k') = f(E_{\text{rel}}, \theta). \tag{4.12}$$

This result provides the justification for the partial wave expansion, an expansion in terms of scattering eigenstates of the operators $\hat{H}_0$, $\hat{l}^2$ and $\hat{l}_z$, in Sect. 1.2. In this basis, the on-shell T-matrix element is diagonal

$$\begin{aligned}
\langle E, l', m'|\hat{\mathsf{T}}|E, l, m\rangle &= \delta_{l,l'}\,\delta_{m,m'}\,\frac{(2l+1)}{2ik}(\text{e}^{2i\delta_l(E)} - 1) \\
&= \delta_{l,l'}\,\delta_{m,m'}\,f_l(E).
\end{aligned}$$

It corresponds to the partial scattering amplitude in (1.22).

---

[4] Scattering with explicitly spin-dependent potentials has been discussed in Sect. 1.4.2 and will be continued in Sect. 4.3.

4.2.3   Mirror Symmetry: Parity

Under the parity operation $\hat{O}_{\text{par}}$ an eigenstate of the position operator changes into a state, which corresponds to the negative direction (likewise for momentum eigenstates)

$$\hat{O}_{\text{par}}|r\rangle = |-r\rangle, \quad \hat{O}_{\text{par}}|k\rangle = |-k\rangle.$$

Further properties are

$$\hat{O}_{\text{par}} = \hat{O}_{\text{par}}^{\dagger} = \hat{O}_{\text{par}}^{-1}, \quad \hat{O}_{\text{par}}^{2} = \hat{1}.$$

A possible phase factor, which could be present in the ordinary space or in the momentum space, can be ignored. However, for the action of the parity operator on a state of a particle with good orbital angular momentum, the phase choice

$$\hat{O}_{\text{par}}|l, m\rangle = (-1)^{l-m}|l, -m\rangle$$

is the most commonly used choice in order to accomodate the behaviour of spherical harmonics with respect to space reflections.

If the potential is mirror symmetric $v(r) = v(-r)$, the Hamiltonian for the relative motion commutes with the parity operator

$$[\hat{H}_{\text{rel}}, \hat{O}_{\text{par}}] = 0,$$

and one finds

$$\hat{O}_{\text{par}}^{\dagger}\hat{S}\hat{O}_{\text{par}} = \hat{S}$$

as well as

$$\langle r'|\hat{S}|r\rangle = \langle -r'|\hat{S}|-r\rangle \quad \text{and} \quad \langle k'|\hat{S}|k\rangle = \langle -k'|\hat{S}|-k\rangle.$$

This result states, that the scattering amplitude is invariant under the conditions stated above

$$f(k \to k') = f(-k \to -k'). \tag{4.13}$$

A *particle* moving in the direction $-k$ is scattered with the same probability in the direction $-k'$ as a particle, which moves in the direction k and is scattered in the direction k'.

An additional remark is: Rotational invariance implies parity invariance for simple systems, as the parity operation corresponds to a rotation by $180°$ about an axis perpendicular to k and k'.

4.2.4 Time Reversal

An operator $\hat{O}_{\text{time}}$, which transforms a state $|\psi\rangle$ into its time-reversed one, is anti-unitary. This follows from the requirement that the action of the operator $\hat{O}_{\text{time}}$ on a position eigenstate does not change that state. If the operator acts on a momentum state, the direction of the momentum is reversed

$$\hat{O}_{\text{time}}|r\rangle = |r\rangle, \quad \hat{O}_{\text{time}}|k\rangle = |-k\rangle . \tag{4.14}$$

This requirement implies that the anti-unitary operator must be anti-linear (compare Sect. 4.1). Each linear combination of states is transformed by an anti-linear operator $\hat{O}_{\text{alin}}$ into a corresponding linear combination with complex conjugate coefficients

$$\hat{O}_{\text{alin}}\left(a_1|\psi_1\rangle + a_2|\psi_2\rangle + \ldots \right) = a_1^* \hat{O}_{\text{alin}}|\psi_1\rangle + a_2^* \hat{O}_{\text{alin}}|\psi_2\rangle + \ldots \tag{4.15}$$

An anti-linear operator with the property (4.15) is norm-preserving, as the products of $\hat{O}_{\text{alin}}^{\dagger}$ with $\hat{O}_{\text{alin}}$ correspond to the unit operator

$$\hat{O}_{\text{alin}}^{\dagger} \hat{O}_{\text{alin}} = \hat{O}_{\text{alin}} \hat{O}_{\text{alin}}^{\dagger} = \hat{1}. \tag{4.16}$$

For instance, the inner product of a state

$$|\psi\rangle = \sum_i a_i |\psi_i\rangle$$

is

$$\langle\psi|\psi\rangle = \sum_{i,k} a_i^* a_k \langle\psi_i|\psi_k\rangle.$$

For an orthonormal basis, $\langle\psi_i|\psi_k\rangle = \delta_{i,k}$, this inner product has the value

$$\langle\psi|\psi\rangle = \sum_i a_i^* a_i.$$

The same value is obtained for

$$\langle\psi|\hat{O}_{\text{alin}}^{\dagger} \hat{O}_{\text{alin}}|\psi\rangle = \sum_{i,k} a_i a_k^* \langle\psi_i|\psi_k\rangle = \sum_i a_i a_i^* .$$

If one compares the expansion of a time-reversed state

$$|\psi_t\rangle = \hat{O}_{\text{time}}|\psi\rangle$$

in ordinary space and in momentum space, one finds with the statements (4.14) and the definition (4.15)

- In *ordinary space:*

$$|\psi_t\rangle = \hat{O}_{\text{time}} \int d^3r \; \langle r|\psi\rangle |r\rangle = \int d^3r \; \langle r|\psi\rangle^* \hat{O}_{\text{time}}|r\rangle$$

$$= \int d^3r \; \langle r|\psi\rangle^*|r\rangle = \int d^3r \; \psi^*(r)|r\rangle$$

and

- in *momentum space:*

$$|\psi_t\rangle = \hat{O}_{\text{time}} \int d^3k \; \langle k|\psi\rangle |k\rangle = \int d^3k \; \langle k|\psi\rangle^* \hat{O}_{\text{time}}|k\rangle$$

$$= \int d^3k \; \langle k|\psi\rangle^*|-k\rangle = \int d^3k \; \psi^*(k)|-k\rangle.$$

If one forms a bra-ket combination in the first case with $\langle r|$ and in the second case with $\langle k|$, one finds

$$\langle r|\psi_t\rangle = \psi_t(r) = \psi^*(r),$$
$$\langle k|\psi_t\rangle = \psi_t(k) = \psi^*(-k).$$

The time-reversed wave function in the space representation is obtained by complex conjugation. In the momentum representation, the momentum is, in addition, replaced by the negative momentum.

The time reversal operator commutes with the Hamiltonian, if the interaction between spinless particles is local and real

$$\hat{O}_{\text{time}} \, \hat{H} = \hat{H} \, \hat{O}_{\text{time}}.$$

On the basis of the antilinearity (4.15) one can therefore show, by expansion of the exponential function, that the commutation relation

$$\hat{O}_{\text{time}} e^{i\hat{H}t/\hbar} = e^{-i\hat{H}t/\hbar} \hat{O}_{\text{time}}$$

is valid for both the Hamiltonian $\hat{H}$ as well as for $\hat{H}_0$. Furthermore the relation

$$\hat{O}_{\text{time}} \hat{\Omega}_\pm = \hat{\Omega}_\mp \hat{O}_{\text{time}} \quad \text{or} \quad \hat{\Omega}_\pm = \hat{O}_{\text{time}}^\dagger \hat{\Omega}_\mp \hat{O}_{\text{time}}$$

is valid, provided the operations 'commutation of operators' and 'formation of the limiting value $t \to \infty$' can be interchanged. Action of $\hat{O}_{\text{time}}$ on the Møller operators interchanges their role.

Finally one finds for the S-matrix operator

$$\hat{O}_{\text{time}}\hat{S} = \hat{S}^\dagger \hat{O}_{\text{time}} \quad \text{or} \quad \hat{S} = \hat{O}_{\text{time}}^\dagger \hat{S}^\dagger \hat{O}_{\text{time}}. \tag{4.17}$$

With the help of (4.17), one can investigate the consequences of time reversal invariance for the S-matrix elements or for the scattering amplitudes in the special case considered here (spinless particles and local interaction). In this case on has in general

$$\begin{aligned}
\langle \psi_a | \hat{S} | \psi_b \rangle &= \langle \psi_a | \hat{O}_{\text{time}}^\dagger \hat{S}^\dagger \hat{O}_{\text{time}} | \psi_b \rangle = \langle \psi_{ta} | \hat{S}^\dagger | \psi_{tb} \rangle^* \\
&= \langle \psi_{tb} | \hat{S} | \psi_{ta} \rangle.
\end{aligned} \tag{4.18}$$

The probability for the transition from ψ_b to ψ_a is as large as for the transition from the time-reversed state ψ_{ta} into the time-reversed state ψ_{tb}. In particular for plane wave states the relation is

$$\langle \mathbf{k}' | \hat{S} | \mathbf{k} \rangle = \langle -\mathbf{k} | \hat{S} | - \mathbf{k}' \rangle.$$

The probability for a process with $\mathbf{k} \to \mathbf{k}'$ is as large as for a process in which the particles move in the opposite direction and their roles are reversed

$$f(\mathbf{k} \to \mathbf{k}') = f(-\mathbf{k}' \to -\mathbf{k}).$$

This property is called reciprocity. If parity invariance is also valid, then the statement

$$f(\mathbf{k} \to \mathbf{k}') = f(\mathbf{k}' \to \mathbf{k})$$

follows from (4.13). This is known as the detailed balance of a reaction process.

4.3 Elastic Scattering: Particles with Spin

If the particles carry a spin, the detailed discussion becomes a little more involved. In this section only the outline of the arguments is presented. Explicit selection rules for the scattering amplitudes for various forms of spin dependence of the interaction potentials in two-particle systems used in practice have already been discussed in Sect. 3.5.2.

A one-particle basis for particles with spin is spanned by states of the form

- Space-spin representation: $|r, \sigma\rangle = |r\rangle \otimes |\sigma\rangle$,
- Momentum-spin representation: $|k, \sigma\rangle = |k\rangle \otimes |\sigma\rangle$.

The associated wave functions correspond to bra-ket combinations with states like

$$|\alpha, s\mu_s\rangle \quad \rightarrow \quad |lm, s\mu_s\rangle = |lm\rangle \otimes |s\mu_s\rangle,$$

so for example

$$\Psi_{l,m,s,\mu_s}(r, \sigma) = \langle r, \sigma|lm, s\mu_s\rangle = \langle r|lm\rangle\langle\sigma|s\mu_s\rangle = \psi_{l,m}(r)\chi_{s,\mu_s}(\sigma).$$

Operations in spin space can be handled in the same fashion as operations in ordinary space or in momentum space.[5]
   The action of rotations in spin space can be treated by the spin part

$$\hat{O}_{rot,s}(\omega) = e^{-i\omega\cdot\hat{s}/\hbar}$$

of the rotation operator

$$\hat{O}_{rot,s}(\omega)|\sigma\rangle = |\sigma_{rot}\rangle.$$

In general one has

$$\hat{O}_{rot,s}(\omega)|\sigma\rangle = \sum_{\sigma'}|\sigma'\rangle\langle\sigma'|\hat{O}_{rot,s}(\omega)|\sigma\rangle,$$

where $\langle\sigma'|\hat{O}_{rot,s}|\sigma\rangle$ is a unitary matrix of dimension $(2s + 1)$. In the case of spin-1/2 particles, the sum runs over the states with spin up ($\sigma' = \uparrow$) and with spin down ($\sigma' = \downarrow$). For a rotation around the z-axis with $\omega = \omega e_z$ one obtains a phase factor, if the operator acts on a state $|s\mu_s\rangle$

$$\hat{O}_{rot,s}(\omega e_z)|s\mu_s\rangle = e^{-i\omega\mu_s}|s\mu_s\rangle.$$

The parity operator commutes with angular momentum operators and for this reason also with spin operators, as these have the same properties (with respect to the commutation relations). Therefore one has

$$\hat{O}_{par}|\sigma\rangle = |\sigma\rangle \text{ and } \hat{O}_{par}|s\mu_s\rangle = |s\mu_s\rangle.$$

---

[5] Spin is an intrinsic property of particles, therefore the behaviour of spin under translations need not be considered.

In the case of time reversal, expressed in more descriptive terms, the sense of rotation of the particle's *inner spinning top* changes. This picture is consistent with the ansatz

$$\hat{O}_{\text{time}}|\sigma\rangle = |-\sigma\rangle \text{ and } \hat{O}_{\text{time}}|s\mu_s\rangle = (-1)^{s-\mu_s}|s-\mu_s\rangle,$$

where the phase $(-1)^{s-\mu_s}$ corresponds to the usual convention.[6]

The statements for the spin of one particle can be extended to the total spin of multi-particle systems. The spin operator of a system of two particles is, as already indicated in Sect. 4.2.2

$$\hat{O}_{\text{rot},S}(\omega) = e^{-i\omega\cdot\hat{S}/\hbar},$$

where $\hat{S} = \hat{s}_1 + \hat{s}_2$ is the total spin of this system. In all equations one only has to replace the particle spin s by the total spin S. The general discussion of two-particle systems with spin thus turns out to be a combination of the statements of Sect. 4.2 for the ordinary space or momentum space combined with the extended statements of this section. Some explicit results, such as the selection rules for the scattering amplitudes for spin-spin or spin-orbit interactions, were given in Sect. 3.5.2. Another necessary extension, the discussion of elastic scattering with *spin-polarised* particles is found in Chap. 6.

4.4 Detailed Calculations for Chap. 4

4.4.1 Translation

In order to prove the statement (4.9), one can assume that the two-particle wave function can be expanded in the form

$$\phi(r_1, r_2) \equiv \langle r_1, r_2|\phi\rangle = \sum_{n_1, n_2} \phi_{n_1}(r_1)\phi_{n_2}(r_2).$$

It is sufficient to perform the proof in one-particle space. One is looking for an operator, defined by the relation

$$\hat{O}_{\text{trans}}(a)|r\rangle = |r+a\rangle,$$

so that the matrix representation

$$\langle r'|\hat{O}_{\text{trans}}(a)|r\rangle = \langle r'|r+a\rangle = \delta(r'-r-a) \tag{4.19}$$

---

[6] Compare C. Itzykson and J. B. Zuber: Quantum Field Theory. McGraw-Hill, New York (1985), p. 244.

can be obtained. The operator $\hat{O}_{\text{trans}}$ has the property that the application of two successive displacements by a and b is characterised by the relations

$$\hat{O}_{\text{trans}}(a)\,\hat{O}_{\text{trans}}(b) = \hat{O}_{\text{trans}}(b)\,\hat{O}_{\text{trans}}(a) = \hat{O}_{\text{trans}}(a+b). \tag{4.20}$$

As the operator $\hat{O}_{\text{trans}}$ preserves orthonormality and the completeness of the basis, it is unitary[7] and can be written in the form

$$\hat{O}_{\text{trans}}(a) = e^{-i\hat{A}(a)}.$$

For the determination of the Hermitian operator $\hat{A}(a)$ one uses (4.20). From the relations

$$\left(\hat{O}_{\text{trans}}(a)\right)^{n} = \hat{O}_{\text{trans}}(na)$$

or

$$\hat{A}(na) = n\hat{A}(a)$$

one finds the relation

$$\hat{A}(a) = a \cdot \hat{B},$$

if one performs a shift by the vector a n times. The operator $\hat{A}$ is proportional to a.

In order to obtain an example for an alternative evaluation, one can restrict the discussion to one spatial dimension and analyse the matrix element

$$\langle x'|x + a\rangle = \langle x'|e^{-ia\hat{B}}|x\rangle,$$

which corresponds to Eq. (4.19). Expansion of the exponential operator in low order yields

$$\langle x'|x + a\rangle = \delta(x' - x) - ia\langle x'|\hat{B}|x\rangle + \ldots.$$

If, on the other hand, a Taylor expansion of the function $\langle x'|x + a\rangle$ with respect to the parameter a is used, one finds

$$\langle x'|x + a\rangle = \delta(x' - x) + a\frac{\partial}{\partial x}\langle x'|x\rangle + \ldots.$$

---

[7] Compare the definition in Sect. 4.1.

By comparing the first order of the two expansions in the quantity a one obtains

$$\langle x'|\hat{B}|x\rangle = i\frac{\partial}{\partial x}\delta(x' - x).$$

For a matrix element of the commutator of $\hat{x}$ with $\hat{B}$ one thus arrives at

$$\langle x'|[\hat{x}, \hat{B}]|\phi\rangle = \int\int dxdx''\{\langle x'|\hat{x}|x\rangle\langle x|\hat{B}|x''\rangle - \langle x'|\hat{B}|x\rangle\langle x|\hat{x}|x''\rangle\}\langle x''|\phi\rangle$$

$$= i\int dx''\{x'[\partial_{x''}\delta(x'' - x')] - [\partial_{x'}\delta(x' - x'')]x''\}\phi(x'')$$

$$= -i\{x'\partial_{x'}\phi(x') - \partial_{x'}[x'\phi(x')]\} = +i\phi(x').$$

The starting point and the last line result, after multiplication with $\hbar$ and rewriting $x' \to x$, in

$$\left[x(-i\hbar\partial_x\phi(x)) + i\hbar\partial_x(x\phi(x))\right] = i\hbar\phi(x).$$

One recognises the commutator $[\hat{x}, \hat{p}_x] = i\hbar$ if one identifies $(-i\hbar\partial_x)$ with the momentum operator $\hat{p}_x$.

To extend the discussion to the three dimensional space, one should replace x by $\boldsymbol{r}$ and ∂_x by ∇. The corresponding expansion of the matrix element (4.19) in three spatial dimensions yields in a few steps the formula given in Sect. 4.2.1.

Literature in Chap. 4

1. M. E. Rose: Elementary Theory of Angular Momentum. J. Wiley, New York (1957); Reprint: Dover Publications, New York (1995)
2. C. Itzykson and J. B. Zuber: Quantum Field Theory. McGraw-Hill, New York (1985).

Elastic Scattering: The Analytical Structure of the S-Matrix

Levinson's theorem, which was referred to in Sect. 1.2.3, points to a possible connection between the bound states and the scattering states of a quantum particle moving in a potential. Since scattering solutions are characterised by real wave numbers and bound states by imaginary wave numbers, one might expect, that a discussion of the scattering problem in the full complex plane of wave numbers—which implies a discussion of the analytic continuation of S-matrix elements—might throw some light on this connection. The method used for this purpose is the investigation of the relations between the solutions of the radial differential equations of the potential problem for different boundary conditions. In addition to the physical solution, which is obtained with the boundary conditions used in Chap. 1, one considers boundary conditions, that lead to solutions, which

- are regular at the origin and are normalised in a specific fashion (regular solutions),
- or show a specific asymptotic behaviour (Jost solutions).

For Jost solutions one can, depending on the properties of the potential function, find regions of the complex wave number plane in which these functions are analytic. By combining the Jost solutions and the regular solutions one can find a set of functions of the wave number, the Jost functions, whose analytic properties can be investigated. The Jost functions in turn can be used to express the partial S-matrix elements in a form, which exhibits definite analytic properties. This representation of the partial S-matrix elements allows a proof of Levinson's theorem as well as insight into additional aspects of collision physics, such as resonance and virtual states.

© Springer-Verlag GmbH Germany, part of Springer Nature 2022   137
R. M. Dreizler et al., *Quantum Collision Theory of Nonrelativistic Particles*,
https://doi.org/10.1007/978-3-662-65591-7_5

5.1   The Regular Solutions of the Potential Scattering Problem

The partial wave expansion of the scattering wave function (compare Sect. 1.2 and 3.5)

$$\langle r | \psi_k^{(+)} \rangle = \left(\frac{2}{\pi} \right)^{1/2} \sum_{lm} i^l \frac{R_l(k, r)}{r} Y_{lm}(\Omega_r) Y_{lm}^*(\Omega_k)$$

leads to a set of differential equations for the radial functions R_l of a central potential problem

$$R_l''(k, r) + \left[k^2 - \frac{l(l+1)}{r^2} - \frac{2m_0}{\hbar^2} v(r) \right] R_l(k, r) = 0. \tag{5.1}$$

The dependence of the radial functions in (5.1) on the wave number k is indicated explicitly in the present notation. With standard boundary conditions one obtains the *physical solution* with the asymptotic form

$$R_l(k, r) \overset{r \to \infty}{\longrightarrow} N_l \sin(kr - \frac{l\pi}{2} + \delta_l(k)) .$$

If one uses the relation (3.56) between the phase shift and the partial S-matrix element, one can also write

$$R_l(k, r) \overset{r \to \infty}{\longrightarrow} A_l \left[e^{-ikr} - (-1)^l S_l(k) e^{ikr} \right], \tag{5.2}$$

where the normalisation factor (not relevant in the following) is

$$A_l = -(i)^{(l-1)} \frac{N_l}{2} e^{-i\delta_l(k)} .$$

As the solutions are not *only* determined by the differential equation, but *also* by the boundary conditions, it is possible to find solutions of (5.1) with a different structure. Of interest in the following discussion are the solutions $\varphi_l(k, r)$, which obey the following boundary conditions:

- The functions $\varphi_l(k, r)$ are regular for $r \longrightarrow 0$.
- The functions have for $r \to 0$ the limiting value

$$\lim_{r \to 0} \frac{\varphi_l(k, r)}{r^{l+1}} = 1.$$

The first condition is a standard boundary condition for the solution of the radial equation (5.1). It entails the fact, that the physical solutions $R_l(k, r)$ and the functions $\varphi_l(k, r)$ are proportional to each other in the neighbourhood of $r = 0$. The second condition fixes the normalisation of the functions $\varphi_l(k, r)$. This has far-reaching consequences. The boundary conditions for the functions $\varphi_l(k, r)$ are in contrast to the boundary conditions applied to the physical solution (5.2), *independent* of the wave number. At this point one can resort to a theorem of Poincaré, which states:[1]

> If a differential equation includes an entire function of a parameter, then any solution, which satisfies boundary conditions, which are independent of this parameter, is an analytic function of this parameter in the entire complex plane.

As k^2 is an entire function,[2] the solutions $\varphi_l(k, r)$ satisfy the conditions of this theorem. These functions have no singularity in the complete complex k-plane. For this reason they are called *regular solutions*. Some properties of these functions are:

- The solution $\varphi_l(k, r)$ is an even function of k

$$\varphi_l(-k, r) = \varphi_l(k, r).$$

This statement follows directly from the invariance of the differential equation (5.1) (with φ_l in the place of R_l) under the transformation $k \to -k$.
- If the complex conjugate differential equation to (5.1) (for φ_l) contains angular momentum values, the potential function and the radial coordinate, which are all real

$$\frac{d^2 \varphi_l^*(k^*, r)]}{dr^2} + \left[(k^*)^2 - \frac{l(l+1)}{r^2} - \frac{2m_0}{\hbar^2} v(r) \right] \varphi_l^*(k^*, r) = 0,$$

then comparison with (5.1) shows, that the function $\varphi_l^*(k^*, r)$, obtained with the same boundary conditions, is equal to the function $\varphi_l(k, r)$

$$\varphi_l^*(k^*, r) = \varphi_l(k, r).$$

It follows, that $\varphi_l^*(k, r) = \varphi_l(k, r)$, if the wave numbers are real. The regular solutions are real for real wave numbers.

---

[1] H. Poincaré, Acta Math. **4**, p. 201 (1884).

[2] Entire functions are analytic in the complete complex plane. Locally they can be represented by a convergent power series. Examples are polynomials or the exponential function and sums or products of these functions.

5.2   The Jost Solutions and the Jost Functions

The regular solutions have simple properties in the complex plane, but they differ from the physical solutions $R_l(k, r)$. In order to establish a relationship between the functions $\varphi_l(k, r)$ and $R_l(k, r)$, it is necessary to consider a third set of solutions of the differential equation (5.1). These solutions,[3] which are termed *Jost solutions* $f_l^{(\pm)}(k, r)$ are determined by the following boundary conditions: The asymptotic limit of the Jost solutions are exponential functions

$$\lim_{r \to \infty} (e^{\pm ikr} f_l^{(\pm)}(k, r)) = 1 \quad \text{or} \quad \lim_{r \to \infty} f_l^{(\pm)}(k, r) = e^{\mp ikr}. \tag{5.3}$$

The functions $f_l^{(+)}(k, r)/r$ correspond (for positive values of the wave number) asymptotically to waves, which move in the direction of the scattering centre at $r = 0$, while the functions $f_l^{(-)}/r$ describe spherical waves emanating from the scattering centre.[4] Concerning the behaviour for $r \to 0$ no conditions are imposed, so that in general the Jost solutions will not regular at the origin. Since the boundary conditions depend on the wave number, the Poincaré theorem does not apply to the Jost solutions. However, it turns out, that they have simpler analytic properties than the physical solutions.

5.2.1   Jost Solutions

Some properties of the Jost solutions can be gleaned directly from the differential equation (5.1) and the boundary conditions. One finds:

- The statement

$$f_l^{(+)}(-k, r) = f_l^{(-)}(k, r) \tag{5.4}$$

  follows, if one replaces k by $-k$. It expresses the invariance of the differential equation with respect to the substitution $k \to -k$.
- If the angular momentum values and the coordinate are real, then complex conjugation of the boundary conditions yields

$$f_l^{(+)}(k^*, r)^* = f_l^{(-)}(k, r), \tag{5.5}$$

---

[3] R. Jost, Helv. Phys. Acta **20**, p. 256 (1947).
[4] Notice: The definition $f_l^{(\pm)}(k, r) \to e^{\pm ikr}$ is also used in the literature. The connection of $f_l^{(\pm)}(k, 0)$ with the partial S-matrix elements, discussed below, must be modified accordingly in this case.

again in conformity with the differential equation (provided the potential is real). If the wave number is real, (5.5) simplifies to $f_l^{(+)}(k, r)^* = f_l^{(-)}(k, r)$.

These relations reflect the analytic properties of the Jost solutions. For example, a certain behaviour of $f_l^{(-)}$ in the lower half-plane leads to a corresponding behaviour of $f_l^{(+)}$ in the upper half-plane. However, the proof of the details of these analytic properties is quite lengthy.[5] The result of the investigation of the analytic properties is:

- If the integral $I_1 = \int_0^\infty r|v(r)|dr$ with the potential function $v(r)$ is bounded, i.e. $I_1 < M_1$, then
  - $f_l^{(-)}(k, r)$ is analytic in k in the upper half-plane ($\text{Im}(k) \geq 0$), except for $k = 0$,
  - $f_l^{(+)}(k, r)$ is analytic in k in the lower half-plane ($\text{Im}(k) \leq 0$), except for $k = 0$.
- If the potential in $I_2 = \int_0^\infty r|v(r)|e^{2pr}dr$ decreases even more rapidly, so that $I_2 < M_2$ with positive $p > 0$, then
  - $f_l^{(-)}(k, r)$ is analytic in k for k-values with $\text{Im}(k) \geq -p$ (the domain is extended into the lower half-plane),
  - $f_l^{(+)}(k, r)$ is analytic in k for k-values with $\text{Im}(k) \leq p$.

These precise statements can be summarised in the simple form: The more rapidly $v(r)$ approaches 0 for $r \to \infty$, the larger is the area of the complex k-plane, in which the functions $f_l^{(\pm)}(k, r)$ are analytic in k.

The two Jost solutions are linearly independent. For a proof, one evaluates the Wronskian (the Wronski determinant) with the asymptotic form and finds

$$W(f_l^{(+)}(k, r), f_l^{(-)}(k, r)) \to f_l^{(+)}(k, r)\frac{\partial f_l^{(-)}(k, r)}{\partial r} - f_l^{(-)}(k, r)\frac{\partial f_l^{(+)}(k, r)}{\partial r}$$

$$= 2ik.$$

For this reason one can represent other solutions of the differential equation (5.1), such as the regular solutions, as linear combinations of the Jost solutions. With the ansatz

$$\varphi_l(k, r) = \frac{1}{2ik}\left[F_l^{(+)}(k)f_l^{(-)}(k, r) - F_l^{(-)}(k)f_l^{(+)}(k, r)\right] \tag{5.6}$$

one introduces the coefficients $F_l^{(\pm)}(k)$, which depend only on the wave number. They are called *Jost functions*. These functions play a central role in the following

---

[5] Interested readers will find a summary of the arguments, for example in R. G. Newton, J. Math. Phys. **1**, p. 319 (1960).

discussion. They have to be chosen, so that the ansatz (5.6) satisfies the boundary conditions $\varphi_l(k, 0) = 0$ at the origin.

5.2.2   Jost Functions

The Jost functions can alternatively be defined via the Wronskian of the Jost solutions and the regular solutions. Using (5.6) one obtains

$$
\begin{aligned}
W(f_l^{(-)}(k, r), \varphi_l(k, r)) &= \frac{1}{2ik} \Big[F_l^{(-)}(k) W(f_l^{(+)}(k, r), f_l^{(-)}(k, r)) \\
&\quad - F_l^{(+)}(k) W(f_l^{(-)}(k, r), f_l^{(-)}(k, r)) \Big] \\
&\longrightarrow \frac{1}{2ik} F_l^{(-)}(k)(2ik) = F_l^{(-)}(k),
\end{aligned}
\tag{5.7}
$$

as well as

$$
W(f_l^{(+)}(k, r), \varphi_l(k, r)) \longrightarrow \frac{1}{2ik} F_l^{(+)}(k)(2ik) = F_l^{(+)}(k).
\tag{5.8}
$$

The properties of the Jost functions are a direct consequence of the properties of the regular solutions and the Jost solutions:

- The statements

$$
\varphi_l(k, r) = \varphi_l(-k, r) \quad \text{and} \quad f_l^{(-)}(k, r) = f_l^{(+)}(-k, r)
$$

for real and complex k-values lead with (5.6) to the result

$$
F_l^{(+)}(k) = F_l^{(-)}(-k).
\tag{5.9}
$$

This symmetry shows that it is actually sufficient to consider one Jost function, for example

$$
F_l(k) \equiv F_l^{(+)}(k)
\tag{5.10}
$$

for each value of the angular momentum. In this case one finds for the second Jost function $F_l^{(-)}(k) = F_l(-k)$. The ansatz (5.6) for the regular solution can also be written in the form

$$
\varphi_l(k, r) = \frac{1}{2ik} \Big[F_l(k) f_l^{(-)}(k, r) - F_l(-k) f_l^{(+)}(k, r) \Big].
\tag{5.11}
$$

- The properties under complex conjugation

$$\varphi_l^*(k^*, r) = \varphi_l(k, r) \quad \text{and} \quad f_l^{(+)}(k^*, r)^* = f_l^{(-)}(k, r)$$

lead to the symmetry relation (details in Sect. 5.5.1)

$$F_l^*(k^*) = F_l(-k) \tag{5.12}$$

for the Jost function $F_l(k)$.

The regular solution can be represented in terms of the Jost solutions $f_l^{(\pm)}(k, r)$ in the form

$$\varphi_l(k, r) = -\frac{F_l(-k)}{2ik} \left[f_l^{(+)}(k, r) - \frac{F_l(k)}{F_l(-k)} f_l^{(-)}(k, r) \right] \tag{5.13}$$

as a consequence of this list of properties of the Jost functions. This form allows the extraction of a representation of the partial S-matrix elements in terms of the Jost function(s). The equal boundary conditions for the regular and physical solutions at the origin require a strict proportionality of the two functions

$$\varphi_l(k, r) = c_l(k) R_l(k, r) \tag{5.14}$$

for all r values, therefore also in the asymptotic region. There one finds

$$\varphi_l(k, r) \xrightarrow{r \to \infty} \frac{iF_l(-k)}{2k} \left[e^{-ikr} - \frac{F_l(k)}{F_l(-k)} e^{ikr} \right]$$

on one hand and

$$R_l(k, r) \xrightarrow{r \to \infty} A_l(k) \left[e^{-ikr} - (-1)^l S_l(k) e^{ikr} \right]$$

on the other (see (5.2)). A comparison of the right- and left-hand sides of (5.14) gives the relation

$$S_l(k) = (-1)^l \frac{F_l(k)}{F_l(-k)}. \tag{5.15}$$

On the basis of the relation (5.15) the properties of the Jost functions can be used to confirm the properties of the S-matrix, which have been found in Sect. 3.3.2 and 3.6.7. They are

$$S_l^*(k^*) S_l(k) = \frac{F_l^{(+)}(k^*)^* F_l^{(+)}(k)}{F_l^{(-)}(k^*)^* F_l^{(-)}(k)} = \frac{F_l^{(-)}(k) F_l^{(+)}(k)}{F_l^{(+)}(k) F_l^{(-)}(k)} = 1. \tag{5.16}$$

This result represents an analytic continuation of the unitarity of the S-matrix. The simpler form of the optical theorem, given in (3.55),

$$S_l^*(k)S_l(k) = 1$$

follows for real k values. In addition one finds with (5.15) directly

$$S_l(k)S_l(-k) = 1 \quad \text{or} \quad S_l^{-1}(k) = S_l(-k), \tag{5.17}$$

that is an equation, which describes the behaviour of the S-matrix under time reversal (naively: $k \to -k$, compare the remarks in Sect. 4.2). The investigation of the analytic structure of the S-matrix has been reduced, with (5.15), to the analytic properties of the Jost functions in the complex wave number plane. A key point of these investigations is the symmetry of the Jost function $F_l(k)$ in the complex k-plane, which is expressed by the relation (5.12)

$$F_l^*(k^*) = F_l(-k).$$

The detailed properties of the partial S-matrix are presented in the next section.

5.3   The Partial S-Matrix Elements

Statements about the analytic properties of the *partial* S-matrix elements in the complex wave number plane can be obtained from the behaviour of the Jost functions as functions of complex k-values. On the basis of Eqs. (5.15) and (5.16), one finds, that knowledge of the values of the S-matrix in one quadrant, allows the calculation of the corresponding values of the S-matrix in the three adjacent quadrants: The value of the S-matrix at position k

$$S_l(k) = S_l$$

determines the values at the positions k^*, $-k^*$ and $-k$, which are symmetric with respect to k.

$$S_l(k) = S_l \quad \to \quad S_l(k^*) = \frac{1}{S_l^*}, \ S_l(-k^*) = S_l^*, \ S_l(-k) = \frac{1}{S_l}.$$

These statements, which are illustrated in Fig. 5.1, establish a symmetry with respect to the position of zeros and poles of the S-matrix for each value of the angular momentum l in the complex k-plane. If there is a zero at the point k, then one will find a pole at the point $-k$ (a corresponding statement would hold for a pole). A pole also exists at the point k^*. Both the zeros and the poles of the S-matrix occur in pairs symmetric with respect to the imaginary k-axis, and as pairs of a zero and a pole, which are symmetric with respect to the real k-axis. If one wishes to examine

Fig. 5.1 Symmetry of the partial S-matrix $S_l(k)$ in the complex k-plane

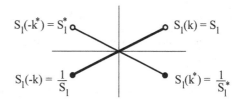

the situation in more detail, it is useful to begin with a discussion of the points on the axes of the complex k-plane. Points, which are not on the axes, will be addressed in the following section.

5.3.1 Scattering States and Bound States

Values of k on the real axis correspond to $E = (\hbar^2 k^2)/(2m_0)$, that is positive and real energy values. This is the domain of scattering states. The representation of the asymptotic regular solution in terms of Jost solutions is

$$\varphi_l(k, r) \xrightarrow{r \to \infty} \frac{1}{2ik} \left[F_l(k)e^{ikr} - F_l(-k)e^{-ikr} \right].$$

The sign of k indicates, after appropriate definition in three-dimensional space, a movement towards or away from the scattering centre. For the S-matrix elements the relation

$$S_l(-k) = 1/S_l(k)$$

is valid. It allows a representation of the S-matrix by the phase shifts δ_l

$$S_l(k) = \exp(2i\delta_l(k)) \qquad \text{with} \qquad \delta_l(-k) = -\delta_l(k).$$

It is also useful to represent the asymptotic limit of the regular wave function, e.g. on the basis of (5.13) and (5.3), for values of the variable k on the imaginary axis

$$k = \pm i\, e_n \quad \text{with real} \quad e_n > 0.$$

The points on the imaginary axis correspond to negative, real energy values

$$E_n = -\frac{\hbar^2}{2m_0} e_n^2.$$

They represent therefore possible discrete bound eigenstates in the potential $v(r)$ or bound states of a two-particle system with a corresponding interaction $v(r) =$

$w(|r_2 - r_1|)$ between the particles. However, not all points of this kind on the imaginary axis correspond to bound states, as will be explained below.

If one considers a point on the positive imaginary axis with $k = +i\,e_n$, one obtains

$$\varphi_l(i\,e_n, r) \longrightarrow \frac{1}{2e_n}\left[F_l(-i\,e_n)e^{e_n r} - F_l(i\,e_n)e^{-e_n r}\right].$$

For the expected asymptotic form of a normalisable state one must have

$$F_l(-i\,e_n) = 0 \quad \text{and} \quad F_l(i\,e_n) \neq 0.$$

In this case, the S-matrix has the value $\pm\infty$

$$S_l(i\,e_n) = (-)^l \frac{F_l(k)}{F_l(-k)} = (-)^l \frac{F_l(i\,e_n)}{F_l(-i\,e_n)} \to \pm\infty.$$

For points on the negative imaginary axis $k = -i\,e_n$ the corresponding statement is

$$\varphi_l(-i\,e_n, r) \longrightarrow -\frac{1}{2e_n}\left[F_l(i\,e_n)e^{-e_n r} - F_l(-i\,e_n)e^{e_n r}\right]. \tag{5.18}$$

In this case, one needs

$$F_l(i\,e_n) \neq 0 \quad \text{and} \quad F_l(-i\,e_n) = 0,$$

so that

$$S_l(-i\,e_n) = (-1)^l \frac{F_l(-i\,e_n)}{F_l(i\,e_n)} = 0.$$

It was shown at the beginning of this chapter, that the occurrence of (with respect to the real axis) symmetrically arranged poles and zeros is a consequence of the condition (5.16). However, the discussion of singularities or zeros on the negative imaginary axis is, as noted by Ma[6] in the form of explicit examples and by Jost[7] in general form, not so simple.

---

[6] S. T. Ma, Phys. Rev. **69**, p. 668 (1946) and **71**, p. 195 (1947).
[7] R. Jost, Helv. Physica Acta **20**, p. 256 (1947).

According to Eq. (5.18) the regular solution for k-values with a vanishing S-matrix element has the asymptotic form

$$\varphi_l(-\mathrm{i}\,e_n, r) \longrightarrow -\frac{1}{2e_n} F_l(\mathrm{i}\,e_n)\mathrm{e}^{-e_n r}.$$

However, it turns out that the magnitude of the Jost function for these k-values is infinite

$$|F_l(\mathrm{i}\,e_n)| = \infty.$$

This means that the function $\varphi_l(-\mathrm{i}\,e_n, r)$ is not normalisable. For this reason, the zeros of the S-matrix on the negative imaginary axis are called *redundant zeros*. One of the explicit examples of the computation of Jost solutions and Jost functions studied by Ma are S-states in an exponential potential. These calculations are presented in Sect. 5.5.2.

In order to present a summary of this discussion, one can assume that the range of the potential is short enough, so that the singularities of the S-matrix are only generated by the zeros of the denominator $F_l^{(-)}(k)$. With this assumption one can prove the following statements (details in Sect. 5.5.3):

- In the upper half-plane $\mathrm{Im}(k) > 0$, one finds only zeros of $F_l^{(-)}(k)$ on the imaginary axis. As the energy for $k = \mathrm{i}e_n$ ($e_n > 0$, real) is negative and real, $E = -(\hbar^2 e_n^2)/(2m_0)$, these poles of the S-matrix elements correspond to bound states.
- There can be no zeros of $F_l^{(-)}(k)$ on the real axis. This is the region of positive energies (scattering states). The unitarity relation

$$S_l(-k) = 1/S_l(k)$$

holds for these S-matrix elements.
- In the entire lower half-plane ($k = a - \mathrm{i}b$, $b > 0$) there is no restriction on the distribution of the zeros of $F_l^{(-)}(k)$. There is, however, according to the relations (5.9) and (5.12),

$$F_l^{(+)}(-k^*)^* = F_l^{(-)}(-k) = F_l^{(+)}(k)$$

or alternatively

$$F_l^{(-)}(k) = F_l^{(+)}(-k) = F_l^{(-)}(-k^*)^*$$

a left-right symmetry of poles and zeros of the S-matrix elements. The physical interpretation of the poles of the S-matrix in the lower complex k-plane outside the imaginary axis is clarified in the next section.

5.3.2   Resonance and Virtual States

Poles in the lower half-plane correspond to *quasi-stationary* states, which characterise the capture of a particle in a or the escape of a particle from a potential region.[8] For a simple explanation of this assertion, one considers the regular solution of the scattering problem, including the time-dependent factor

$$\Phi_l(k; r, t) = \frac{1}{r}\varphi_l(k, r) \exp\left[-\frac{i}{\hbar}E(k)t\right]$$

at points of the lower half-plane

$$k \equiv k_{\text{res}} = \pm a - ib \qquad \text{with} \quad a > 0, \ b > 0.$$

The energy value

$$E_{\text{res}} = \frac{\hbar^2}{2m_0}k_{\text{res}}^2 = \frac{\hbar^2}{2m_0}\left\{[a^2 - b^2] \mp \frac{i}{2}[4ab]\right\}$$

$$= \bar{E}_{\text{res}} \mp \frac{i}{2}\Gamma$$

$$(5.19)$$

leads to a time-dependent probability distribution for a particle at a distance r from the origin

$$|\Phi_l(k_{\text{res}}; r, t)|^2 = \frac{1}{r^2}|\varphi_l(k_{\text{res}}, r)|^2\exp\left[\mp\frac{\Gamma}{\hbar}t\right].$$

The time factor states:

- If the particle is in the third quadrant (positive sign), the probability increases with time. Trapping of particles takes place.
- If the particle is in the fourth quadrant (negative sign), the probability decreases with time. The *state* decays.

The average lifetime of these states is determined by the formula $\tau = \hbar/\Gamma$. According to the definition $\Gamma \propto 4ab$ in (5.19), one finds, for a given real part of the wave number, that the lifetime increases with decreasing imaginary part.

---

[8] Alternatively, in the case of a two-particle system: the formation of a temporary bound state and the decay of such a state.

The space component of the wave function characterising a pole of the S-matrix in the lower half-plane has the asymptotic form

$$\frac{1}{r}\varphi_l(k, r) \xrightarrow{r \to \infty} \frac{1}{r} e^{\pm iar} e^{br},$$

i.e. the form of an incoming (outgoing) spherical wave in the third (fourth) quadrant, which is multiplied in each case by a factor increasing with distance.

For a more precise characterisation of the quasi-stationary states, one may consider the S-matrix in the form

$$S_l(k) = (-1)^l \frac{F_l^{(+)}(k)}{F_l^{(-)}(k)} = (-1)^l \frac{F_l^{(-)}(k^*)^*}{F_l^{(-)}(k)}$$

and expand the Jost function $F_l^{(-)}(k)$ in a power series, for example, around a pole at $k_{\text{res}} = a - ib$

$$F_l^{(-)}(k) = A_l(k_{\text{res}})(k - k_{\text{res}}) + \ldots.$$

The variation of $S_l(k)$ as a function of real k-values is the quantity that is investigated in scattering experiments. In lowest order, one finds for the S-matrix in the vicinity of the pole the expansion

$$S_l(k) = (-1)^l \frac{A_l^*}{A_l} \cdot \frac{(k - k_{\text{res}}^*)}{(k - k_{\text{res}})}.$$

For points on the real axis with $b = 0$ the S-matrix can be represented by a phase shift (see Eq. (3.56))

$$S_l(k) = (-1)^l \frac{A_l^*}{A_l} = \exp(2i\delta_l) \quad \text{for} \quad b = 0.$$

For points in the lower half of the complex k-plane the one-pole expansion of the Jost functions leads therefore to

$$S_l(k) = \exp(2i\delta_l) \cdot \frac{(k - k_{\text{res}}^*)}{(k - k_{\text{res}})}. \tag{5.20}$$

The contribution of $S_l(k)$ to the partial cross section is given by (3.57)

$$\sigma_l = \frac{\pi}{k^2}(2l + 1)|1 - S_l(k)|^2.$$

Fig. 5.2 Lorentz curve

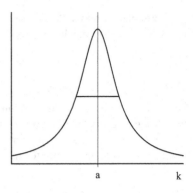

After a slightly tedious evaluation of the expression $|1 - S_l(k)|^2$

$$|1 - S_l(k)|^2 = (1 - \text{Re}[S_l(k)])^2 + (\text{Im}[S_l(k)])^2,$$

one obtains the result

$$|1 - S_l(k)|^2 = 4\frac{((k-a)^2 - b^2)}{((k-a)^2 + b^2)}\sin^2\delta_l - \frac{8b(k-a)}{((k-a)^2 + b^2)}\sin\delta_l\cos\delta_l$$
$$+ \frac{4b^2}{((k-a)^2 + b^2)}.$$

In the first term one recognises a contribution, which corresponds to a modified potential scattering. The last term is a direct consequence of the pole of the scattering matrix, in addition there appears an interference term. The pole contribution has the form of a Lorentz curve (Fig. 5.2). The maximum at the position $k = a$ has the value 4, the half-width is $\Delta = 2b$. It has a typical resonance structure, if b is small (compare Dreizler/Lüdde, Vol. 1, Sect. 4.4.2). The resonance structure has a noticeable effect on the cross section only if the pole is not too far away from the real axis. The modification and the additional terms vanish on the real axis ($b = 0$).

The *one-pole approximation* of the S-matrix, defined in (5.20), satisfies the unitarity condition (5.16)

$$S_l(k)^* S_l(k) = 1$$

for real k-values, but not the condition (5.17) for reflection symmetry with respect to the real axis

$$S_l(k) S_l(-k) = 1.$$

However, the symmetry can be restored by inclusion of the zero associated with the pole in the upper half-plane.

Fig. 5.3 Potential pocket

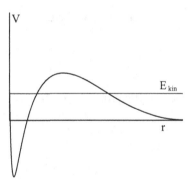

Figure 5.3 illustrates a situation, for which a resonance can be observed. The combination of a central potential with the centrifugal potential

$$v_{\text{eff}}(r) = v_c(r) + \frac{\hbar^2 \, l(l+1)}{2m_0 \, r^2}$$

can generate a *potential pocket*. An incident particle, whose kinetic energy is less than the height of the barrier, is either reflected directly (this is the contribution of the background phase $\exp[2i\delta_l]$) or it can tunnel through the barrier. If the particle remains in a quasi-stationary state inside the pocket for a period of time before it leaves the pocket by tunneling, a resonant contribution to the cross section results.

Poles of the S-matrix on the negative imaginary axis turn into a double pole as a result of the left-right symmetry. The regular solution has the asymptotic form

$$\varphi(k, r) \overset{r \to \infty}{\longrightarrow} e^{ik_{\text{res}} r} \to e^{br}, \qquad k_{\text{res}} = -ib, \; b > 0$$

in this case. This wave function can not be normalised. These states with the energy $E = -\hbar^2 b^2 / 2m_0$ are referred to as *virtual states*. The associated poles can also affect the cross sections. If such a state is near the real axis, the single-pole approximation for the S-matrix gives

$$S_l(k) \propto \frac{k - ib}{k + ib}.$$

The contribution to the cross section is

$$\frac{4b^2}{k^2 + b^2}.$$

It does not vanish in the limit $k \to 0$.

5.3.3   Levinson's Theorem

Levinson's theorem, as described in Sect. 1.2, connects the number of bound states in a potential (of finite depth) with the difference of the phase shifts for $k = 0$ and $k \to \infty$. This is a consequence of the analytic structure of the S-matrix, or rather of the analytic properties of the Jost functions. The starting point of the discussion of the theorem is the partial S-matrix for real values of the wave number, which can be written in either the form

$$S_l(k) = e^{2i\delta_l(k)} \qquad \text{or} \qquad S_l(k) = (-1)^l \, \frac{F_l(k)}{F_l(-k)}.$$

The proof of the theorem requires two steps (additional notes concerning the proof can be found in Sect. 5.5.4):

- The logarithmic derivative of the first form gives

$$\frac{d \ln S_l(k)}{dk} = 2i \frac{d \, \delta_l(k)}{dk}.$$

Calculation of the integral

$$I = \int_{-\infty}^{\infty} dk \, \frac{d \ln S_l(k)}{dk} = 2i \int_{-\infty}^{\infty} dk \, \frac{d \, \delta_l(k)}{dk}$$

gives

$$I = 4i \int_0^{\infty} dk \, \frac{d \, \delta_l(k)}{dk} = 4i \, [\delta_l(\infty) - \delta_l(0)],$$

as the phase shifts are odd functions of k.
- With the second form one evaluates the logarithmic derivative of the S-matrix

$$\frac{d \ln S_l(k)}{dk} = \frac{d}{dk} \ln \left[(-1)^l \, \frac{F_l(k)}{F_l(-k)} \right] = \frac{d}{dk} \, [\ln F_l(k) - \ln F_l(-k)] .$$

In the integral

$$I = \int_{-\infty}^{\infty} dk \, \frac{d}{dk} \, [\ln F_l(k) - \ln F_l(-k)]$$

one can use the substitution $k \to -k$ in the second term and find

$$I = 2 \int_{-\infty}^{\infty} dk \, \frac{d}{dk} \ln F_l(k).$$

Fig. 5.4 Contour for the
proof of Levinson's theorem

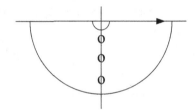

The Jost function $F_l(k)$ goes to zero for $|k| \to \infty$ in the lower half-plane. Therefore, one can recast the integral I into a contour integral (Fig. 5.4) with an infinite, negatively oriented semicircle in the lower half-plane

$$I = 2 \int_C dk \frac{d}{dk} \ln F_l(k).$$

According to the theorem of residues, each isolated zero of

$$F_l(k) = F(-ie_n) \qquad \text{for} \quad e_n > 0$$

gives a contribution with residue 1 at each of the finite number of points of the lower half-plane, which correspond to bound states with angular momentum l within the contour. For this reason one has

$$I = -4\pi i N_l(\text{bound})$$

and

$$[\delta_l(0) - \delta_l(\infty)] = +\pi N_l(\text{bound}). \tag{5.21}$$

Usually one defines the phase shifts in such a way, that $\delta_l(\infty)$ is equal to zero. A modification is necessary for angular momentum $l = 0$, namely if the Jost function $F_0(k)$ has the value zero for $k = 0$. It can be shown[9] that this zero is simple (of order 1) and that the Jost function behaves exactly as

$$F_0(k) \longrightarrow k \quad \text{for} \quad k \to 0.$$

In this case one has to circumvent the origin of the k-plane by a small semicircle (radius ϵ with $\epsilon \to 0$) and obtain an additional contribution

$$\Delta I = -2\pi i,$$

---

[9] R. G. Newton: J. Math. Phys, **1**, p. 319 (1960).

so that (5.21) must be replaced by

$$[\delta_0(\infty) - \delta_0(0)] = -\pi \left(N_0(\text{bound}) + \frac{1}{2} \right)$$

for $l = 0$.

5.4   Illustration of the Theorem: The Spherical Potential Well

The remarks of the last sections can be illustrated by the example of the spherical potential well, for simplicity for states with $l = 0$ (compare Sect. 1.2.3.)

$$v(r) = \begin{cases} \pm v_0 \text{ for} & r \leq R, \\ 0 \quad \text{ for} & r > R \end{cases} .$$

If one uses $v_0 > 0$ the positive sign describes a barrier, the negative sign a well. For the effective wave number K in the potential region one has

$$K^2 = k^2 - U(r) = k^2 \mp \frac{2\,m_0}{\hbar^2} v_0 = k^2 \mp U_0.$$

The appropriate Schrödinger equation is a Bessel-Riccati differential equation

$$R_0''(r) + K^2 R_0(r) = 0.$$

The Jost solutions with the boundary conditions (5.3) can, however, be represented more directly via complex exponential functions

$$f_0^{(\mp)}(k, r) = \begin{cases} A^{(\mp)} e^{iKr} + B^{(\mp)} e^{-iKr}, & r \leq R, \\ e^{\pm ikr}, & r > R. \end{cases}$$

For the coefficients one obtains with the conditions at $r = R$

$$A^{(-)}(k) = \frac{1}{2} \left(1 + \frac{k}{K} \right) e^{i(k-K)R}, \qquad B^{(-)}(k) = \frac{1}{2} \left(1 - \frac{k}{K} \right) e^{i(k+K)R},$$

$$A^{(+)}(k) = \frac{1}{2} \left(1 - \frac{k}{K} \right) e^{-i(k+K)R}, \qquad B^{(+)}(k) = \frac{1}{2} \left(1 + \frac{k}{K} \right) e^{-i(k-K)R}.$$

The solutions satisfy the properties (5.4) and (5.5). The Jost functions are, for example, obtained from the relation (5.6) at $r = 0$

$$F_0(k) \equiv F_0^{(+)}(k) = A^{(+)}(k) + B^{(+)}(k) = e^{-ikR} \left(\cos KR + i\frac{k}{K} \sin KR \right).$$

This information allows the construction of the regular solution and all other relations of the previous sections. Of particular interest, however, is the S-matrix

$$S_0(k) = \frac{F_0(k)}{F_0(-k)} = e^{2ikR} \frac{K \cot KR + ik}{K \cot KR - ik}.$$

The exponential prefactor represents the part, which is generated by the scattering at the step at $r = R$. Simple poles of the S-matrix are defined by the relation

$$K \cot KR = ik, \tag{5.22}$$

simple zeros by

$$K \cot KR = -ik.$$

5.4.1 The Potential Well

For a given set of parameters v_0 and R one can calculate the bound states (poles on the positive imaginary axis) because of $k = i\kappa$, $\kappa > 0$, from the equations

$$KR \cot KR = -\kappa R, \qquad K^2 R^2 = U_0 R^2 - \kappa^2 R^2,$$

or with the quantities $x = KR$ and $y = \kappa R$ as intersections of the curves $y = -x \cot x$ with the circles $x^2 + y^2 = U_0 R^2$ (see Sect. 1.2.3). Depending on the size of the parameter $U_0 R^2$, zero, one, two etc. bound states (Fig. 5.5) can be found.

Poles on the negative imaginary axis correspond to virtual states. They are defined by the positive wave number $\bar{\kappa}$ with $k = -i\bar{\kappa}$ or by the intersection of the curves $y = x \cot x$ with the circles $x^2 + y^2 = U_0 R^2$, now with $y = -\bar{\kappa} R$. Fig. 5.6 illustrates the possible intersections of the two curves. One recognises,

Fig. 5.5 Determination of bound states

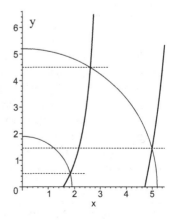

Fig. 5.6 Determination of
virtual states

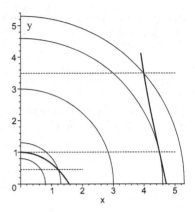

for example, that no virtual state exists for circles with the parameter combination $s = \sqrt{U_0}R < 1$. One state does, however, exist for $1 < s < \pi/2$. For $s > \pi/2$ up to a value for the circle, which just touches the second branch of the cotangent-like curve, again no virtual state is found.

Quasi-stationary states are also given by (5.22), for example, for a decaying state in the fourth quadrant

$$K \cot K R = ik \quad \text{with} \quad k = a - ib, \ (a, b > 0, \text{ real}).$$

In order to determine the complex wave numbers, one resorts to numerical methods or approximations. This point will, however, only be illustrated by the presentation of some results.

An interesting study of the distribution of the poles of the S-matrix for the spherical potential well is given in a paper by H. M. Nussenzveig.[10] It is concerned with the change of the position of the poles by variation of the potential parameters, for example, the depth of the potential well at fixed radius. For $U_0 = 0$ one finds poles at

$$(kR)_n = n\pi - i\infty, \qquad n = 0, \pm 1, \pm 2, \ldots$$

via the solution of the equation $\cot kR = i$. They characterise quasi-stationary states, symmetrical with respect to the imaginary axis, and one virtual state. With increasing U_0 the poles move more or less along a straight line $k = n\pi$ (Fig. 5.7). The pole with $n = 0$ approaches the origin first, crosses the real axis, and becomes a bound state. As the well grows deeper, the poles with $n = \pm 1$ swing onto the straight line with $k = -i$. Shortly before reaching the point $(0, -i)$, the left pole turns upward and moves towards the origin. After crossing the origin, this pole corresponds to a bound state. The right pole in the complex k-plane turns downward

---

[10] H. M. Nussenzveig, Nucl. Phys. **11**, p. 499 (1959).

Fig. 5.7 Shift of the pole
positions with U_0

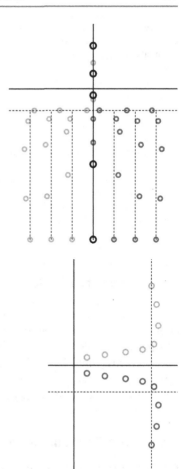

Fig. 5.8 Pole migration for
$l > 0$

and follows the negative imaginary axis. It becomes a virtual state. The poles with
$n = \pm 2, \pm 3, \dots$ follow this pattern with a corresponding delay. For the case with
$l = 0$ no centrifugal barrier exists. This implies that no resonance effects can be
present.

A corresponding calculation and discussion is also possible, but more complicated for $l > 0$. The difference to the case $l = 0$ is the fact, that the wandering
poles do not move on a parallel to the real axis for sufficiently deep potentials,
but (depending on l) approach that axis. True resonances occur, which are formally
described by the combination of a pole near the axis (Fig. 5.8) in the lower half-plane
and a zero in the upper half-plane—which corresponds in reality to the occurrence
of a potential pocket.

Fig. 5.9 Motion of the poles
and zeros in the case of the
potential barrier

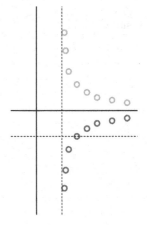

5.4.2   The Potential Barrier

For a potential barrier one finds (for the same initial situation) that the poles move
outward as the height of the barrier increases and, together with the zero in the upper
half-plane, approach the real axis (Fig. 5.9). This suggests again the possibility of
resonance structures, but their mechanism is different: the barrier is permeable for
certain energy values at a certain width. Particles are trapped for a certain period of
time before this quasi-stationary intermediate state decays again.

5.5   Detailed Calculations for Chap. 5

5.5.1   Complex Conjugation of the Jost Functions

It is sufficient to rewrite the representations of the functions $\varphi_l(k^*, r)^*$ and $\varphi_l(k, r)$
according to the definition (5.6) and to compare them. From

$$\varphi_l(k, r) = -\frac{i}{2k} \left[F_l^{(+)}(k) f_l^{(-)}(k, r) - F_l^{(-)}(k) f_l^{(+)}(k, r) \right]$$

follows

$$\varphi_l(k^*, r)^* = \frac{i}{2k} \left[F_l^{(+)}(k^*)^* f_l^{(-)}(k^*, r)^* - F_l^{(-)}(k^*)^* f_l^{(+)}(k^*, r)^* \right]$$

and with the properties of $f_l^{(\pm)}(k^*, r)^*$ in (5.5)

$$\varphi_l(k^*, r)^* = \frac{\mathrm{i}}{2k}\left[F_l^{(+)}(k^*)^* f_l^{(+)}(k, r) - F_l^{(-)}(k^*)^* f_l^{(-)}(k, r)\right].$$

From $\varphi_l(k, r) = \varphi_l(k^*, r)^*$ follows by comparison of the coefficients

$$F_l^{(+)}(k^*)^* = F_l^{(-)}(k)$$

or

$$F_l(k^*)^* = F_l(-k).$$

5.5.2 The Jost Solution $f_0^{(-)}(k, r)$ and the Jost Function $F_0^{(-)}(k)$ for the Exponential Potential

The differential equation for a Jost solution $f(k, r)$ with $l = 0$ is

$$\frac{\mathrm{d}^2 f(k, r)}{\mathrm{d}r^2} + k^2 f(k, r) - \frac{2m_0}{\hbar^2} v(r) f(k, r) = 0$$

for an attractive exponential potential

$$v(r) = -\frac{\hbar^2 v_0}{2m_0}\mathrm{e}^{-ar},$$

$$\frac{\mathrm{d}^2 f(k, r)}{\mathrm{d}r^2} + k^2 f + v_0 \mathrm{e}^{-ar} f = 0.$$

The substitution

$$x = A\mathrm{e}^{-ar/2}$$

with

$$x' = -\frac{a}{2}x, \qquad x'' = \frac{a^2}{4}x$$

offers itself. The second derivative

$$\frac{\mathrm{d}^2 f}{\mathrm{d}r^2} = \frac{\mathrm{d}f}{\mathrm{d}x}(x'') + \frac{\mathrm{d}^2 f}{\mathrm{d}x^2}(x')^2$$

and the choice

$$A = \frac{2\sqrt{v_0}}{a}$$

lead to the transformed differential equation

$$\frac{d^2 f}{dx^2} + \frac{1}{x}\frac{df}{dx} + \left(1 + \frac{\kappa^2}{x^2}\right) f = 0 \quad \text{with} \quad \kappa = \frac{2k}{a}.$$

If one sets $\kappa = i\nu$, one recognises the differential equation of the Bessel functions (see Sect. 1.5.4)

$$\frac{d^2 f}{dx^2} + \frac{1}{x}\frac{df}{dx} + \left(1 - \frac{\nu^2}{x^2}\right) f = 0.$$

The general solution is

$$f_\nu(x) = a J_\nu(x) + b J_{-\nu}(x),$$

as long as ν is not an integer ($\nu \neq n$).

The Jost solution $f_0^{(-)}(k, r)$ should satisfy the boundary condition

$$f_0^{(-)} \to e^{ikr} \qquad \text{for} \qquad r \to \infty.$$

The variable x goes to zero for $r \to \infty$. If one inserts the original parameters, one finds with $\nu = \mp i\kappa$ and the limiting value of the Bessel function

$$J_{\pm i\kappa}(x) \to \frac{1}{\Gamma(\pm i\kappa + 1)} \left(\frac{x}{2}\right)^{\pm i\kappa} = \frac{1}{\Gamma(\pm i\kappa + 1)} \left(\frac{A}{2}\right)^{\pm i\kappa} \left(e^{-ar/2}\right)^{\pm i\kappa}$$

$$= \frac{1}{\Gamma(\pm i\kappa + 1)} \left(\frac{A}{2}\right)^{\pm i\kappa} e^{\mp ikr}.$$

The (normalised) function

$$f_0^{(-)}(k, r) = \left(\frac{A}{2}\right)^{+i\kappa} \Gamma(1 - i\kappa) J_{-i\kappa}(x)$$

with

$$A = \frac{2\sqrt{v_0}}{a} \qquad \kappa = \frac{2k}{a} \qquad x = A e^{-ar/2}$$

is the desired Jost solution with the asymptotic behaviour

$$f_0^{(-)}(k, r) \xrightarrow{r \to \infty} e^{ikr}.$$

The corresponding Jost function is

$$F_0^{(-)}(k) = f_0^{(-)}(k, 0) = \left(\frac{\sqrt{v_0}}{a}\right)^{\frac{2ik}{a}} \Gamma(1 - \frac{2ik}{a}) J_{-2ik/a}\left(\frac{2\sqrt{v_0}}{a}\right).$$

Suggestion: Calculate $f_0^{(+)}(k, r)$ as well as $F_0^{(+)}(k)$ for the exponential potential and discuss the result. Calculate the Jost function for a repulsive exponential potential and compare.

5.5.3 Distribution of the Zeros of the Jost Function $F_l^{(-)}(k)$

The proof of the list of zeros of $F_l^{(-)}(k)$, respectively the poles of the S-matrix, involves the following points:

- Zeros of $F_l^{(-)}(k)$ in the upper half plane $\text{Im}(k) > 0$ can only lie on the imaginary axis.

  Proof: If there is a zero of $F_l^{(-)}(k)$, the regular solution is

$$\varphi_l(k, r) = -\frac{i}{2k} F_l^{(+)}(k) f_l^{(-)}(k, r).$$

From the Schrödinger equation

$$\varphi_l''(k, r) + \left[k^2 - \frac{l(l+1)}{r^2} - \frac{2m_0}{\hbar^2} v(r)\right] \varphi_l(k, r) = 0$$

and the complex conjugate differential equation, one obtains (for real l and r) the statement

$$\varphi_l(k, r)\varphi_l''(k^*, r)^* - \varphi_l''(k, r)\varphi_l(k^*, r)^* = ((k^*)^2 - k^2)\varphi_l(k, r)\varphi_l(k^*, r)^*.$$

The left-hand side of this equation represents the derivative of the Wronskian of the functions φ_l and φ_l^*

$$\frac{d}{dr} W(\varphi_l, \varphi_l^*) = ((k^*)^2 - k^2)\varphi_l(k, r)\varphi_l(k^*, r)^*.$$

Integration of this relation yields

$$\int_0^\infty dr \left(\frac{dW}{dr}\right) = W(r)\Big|_0^\infty = \left(\varphi_l(k,r)\varphi_l'(k^*,r)^* - \varphi_l'(k,r)\varphi_l(k^*,r)^*\right)_0^\infty$$

$$= ((k^*)^2 - k^2)\int_0^\infty dr\; \varphi_l(k^*,r)^*\varphi_l(k,r).$$

For a zero of $F_l^{(-)}(k)$ in the upper half-plane with

$$k = a + ib, \qquad b > 0$$

one finds (as the boundary condition corresponds to the condition of regularity)

$$\varphi_l(k,r)\xrightarrow{r\to 0}0 \qquad \text{and}$$

$$\varphi_l(k,r)\xrightarrow{r\to\infty} \text{const}_1\; e^{i(a+ib)r} = \text{const}_2\; e^{-br} \to 0.$$

Corresponding statements are valid for φ_l^*. As a consequence one can note:
– The function φ_l is square-integrable, since it is analytic and vanishes at the boundaries

$$\int_0^\infty dr\varphi_l(k^*,r)^*\varphi_l(k,r) = A_l \neq 0.$$

– The Wronskian vanishes (at the boundaries).
  If one substitutes the value of k, one is left with the statement

$$-4iab A_l = 0.$$

As b and A_l are not equal to zero, a is equal to zero. The zeros of $F_l^{(-)}(k)$ lie in the upper half-plane on the imaginary axis.
• On the real k-axis no zeros of $F_l^{(+)}(k)$ exist. For real values of $k = k^*$ the relation

$$F_l^{(-)}(k)^* = F_l^{(+)}(k)$$

is valid. For this reason the equation $F_l^{(-)}(k) = 0$ follows from $F_l^{(+)}(k) = 0$ and also $\varphi_l(k,r) \equiv 0$.
An exception can be the point with $k = 0$. For $k = 0$ the two Jost solutions are not linearly independent, and the ansatz

$$\varphi_l(k,r) = c_1 f_l^{(+)}(k,r) + c_2 f_l^{(-)}(k,r)$$

for the regular solution is not adequate. An explicit discussion of the situation at hand is required.

5.5.4   Material for the Proof of Levinson's Theorem

The behaviour of the Jost solutions for $r \to 0$ and $r \to \infty$ is determined by the differential equation

$$f_l''(k, r) + \left[k^2 - \frac{l(l+1)}{r^2} - \frac{2m_0}{\hbar^2} v(r) \right] f_l(k, r) = 0$$

and the boundary condition

$$\lim_{r \to \infty} (e^{\pm ikr}(k, r) f_l^{(\pm)}(k, r)) = 1.$$

The potential term can be neglected (up to exceptions) in comparison with the centrifugal term in the two limits. The function $f_l^{(+)}(k, r)$ is in this case determined by a Bessel-Riccati differential equation. As a result of the boundary condition for $r \to \infty$ the desired solution in the limiting cases is a Hankel-Riccati function[11] with

$$f_l^{(+)}(k, r) \propto w_l^{(-)}(k, r) \quad \left\{ \begin{array}{l} \xrightarrow{\ r \to \infty\ } e^{-ikr} \\[2ex] \xrightarrow{\ r \to 0\ } \dfrac{1}{(kr)^l} \end{array} \right. .$$

From this one finds for the Jost function

$$F_l(k) \xrightarrow{\ r \to 0\ } (l+1) r^l f_l^{(+)}(k, r) \propto \frac{1}{k^l}.$$

The integrand of the contour integral can be given in the form

$$\frac{d}{dk} \ln F_l(k) = \frac{F_l'(k)}{F_l(k)}.$$

One recognises for $1 < s < \pi/2$ once more the behaviour for large k values

$$[\ln F_l(k)]' = \frac{F_l'(k)}{F_l(k)} \to 1/k,$$

---

[11] Abramowitz/Stegun p. 437.

on the other hand one can apply directly a theorem[12] for holomorphic functions of Rouché

$$I = 2 \int_C dk \, \frac{F_l'(k)}{F_l(k)} = -2 \, (2\pi \, i \, N_{\text{bound}}).$$

For the calculation of the line integral over the small semicircle, which results from the exclusion of the origin of the coordinate system, one uses the statement that $F_0(k)$ for $k \to 0$ behaves as $F_0(k) \to k$. One finds therefore the correction

$$\Delta I = 2 \int_{\text{sc}} dk \, \frac{d}{dk} \ln k,$$

as the sense of revolution is negative and with the substitution $k = \epsilon e^{i\phi}$ the final result

$$\Delta I = 2 \int_{\text{sc}} dk \, \frac{1}{k} = 2i \int_\pi^0 d\phi = -2\pi i.$$

5.5.5   The Complex Equation $K \cot KR = ik$

The equation $K \cot KR = ik$ goes over into the equation $\cot kR = i$, if there is no potential present. Replacement of the sine and cosine functions in terms of complex exponential functions on the left-hand side, leads to

$$\frac{e^{ikR} + e^{-ikR}}{e^{ikR} - e^{-ikR}} = 1.$$

Resolution results in

$$e^{-ikR} = -e^{-ikR} \quad \text{or} \quad e^{-ikR} = 0.$$

This demands (for a fixed R) $k = -i\infty$. Due to the periodicity of the cotangent function, the complete solution of the equation $\cot kR = i$ is

$$k = n\pi - i\infty \quad \text{with} \quad n = 0, \pm 1, \pm 2, \ldots.$$

---

[12] See, for example, K. Knopp: Theory of Functions. Dover Publications, New York (1996), Vol. 2, p. 111.

Literature in Chap. 5

1. H. Poincaré, Acta Math. **4**, p. 201 (1884)
2. R. Jost, Helv. Physica Acta **20**, p. 256 (1947)
3. R. G. Newton, J. Math. Phys. **1**, p. 319 (1960)
4. S. T. Ma, Phys. Rev. **69**, p. 668 (1946) and **71**, p. 195 (1947)
5. H. M. Nussenzveig, Nucl. Phys. **11**, p. 499 (1959)
6. K. Knopp: Theory of Functions. Dover Publications, New York (1996)

Elastic Scattering with Spin-Polarised Particles   6

It is possible to prepare particle beams and targets in such a way that the spin vectors of the particles are aligned with respect to a selected direction in space. Experiments with polarised beams and/or with polarised targets provide additional insight into the properties of quantum collision systems. The simplest possibilities

- Spin-1/2 particles scatter from spin-0 particles (or exchanging beam and target, spin-0 particles scatter from spin-1/2 particles),
- spin-1/2 particles scatter from spin-1/2 particles

are discussed in this chapter.

In a scattering experiment, one spatial direction stands out, the beam direction. This is usually chosen as the z-axis. If the beam particles carry a spin s, this spin can be quantised with respect to the beam direction or with respect to a direction perpendicular to it. This direction is called the polarisation direction. It can be defined, in the case of spin-1/2 particles (scattering from spin-0 particles), in the following fashion:

The two-spinor

$$\chi = c_1 \begin{pmatrix} 1 \\ 0 \end{pmatrix} + c_2 \begin{pmatrix} 0 \\ 1 \end{pmatrix} = \begin{pmatrix} c_1 \\ c_2 \end{pmatrix}$$

is used to construct the spin density matrix $[\varrho]$

$$[\varrho] = \left[\chi \chi^\dagger \right] = \begin{pmatrix} c_1 c_1^* & c_1 c_2^* \\ c_2 c_1^* & c_2 c_2^* \end{pmatrix}.$$

© Springer-Verlag GmbH Germany, part of Springer Nature 2022
R. M. Dreizler et al., *Quantum Collision Theory of Nonrelativistic Particles*,
https://doi.org/10.1007/978-3-662-65591-7_6

The density matrix is the basis for the definition of a polarisation vector $\boldsymbol{P}$ by the relation

$$[\varrho] = \frac{1}{2} \left\{ \begin{pmatrix} 1 & 0 \\ 0 & 1 \end{pmatrix} + \boldsymbol{P} \cdot [\boldsymbol{\sigma}] \right\}.$$

The vector $[\boldsymbol{\sigma}]$ stands for a vector constructed from the three Pauli matrices. The alternative form

$$\boldsymbol{P} = \langle \boldsymbol{\sigma} \rangle = \mathrm{tr}\,[[\varrho][\boldsymbol{\sigma}]]$$

can be derived from the equation defining $\boldsymbol{P}$. It identifies the polarisation vector $\boldsymbol{P}$ as a *vector*, which points in the direction of the *spin vector* and allows its construction from the components of a spinor.

In order to examine the differential cross section, the scattering amplitudes for the individual transitions are combined into a scattering matrix

$$[F] = \begin{pmatrix} f_{11} & f_{12} \\ f_{21} & f_{22} \end{pmatrix}.$$

This allows to write the expression for the differential cross section in matrix form

$$\frac{d\sigma}{d\Omega} = \left[[\varrho][F]^\dagger[F] \right],$$

which permits the calculation of the cross sections for arbitrary initial situations (for example, in a beam with mixed spin orientations: p % polarised in the x-direction, $(1 - p)$ % in the y-direction).

The collision system spin-1/2 particles scatter from spin-1/2 particles is treated in the same way, except that the spin density matrix and the scattering matrix are (4×4)-matrices, which corresponds to the 16 spin-to-spin channels

$$
\begin{aligned}
(++) &\longrightarrow (++),\ (+-),\ (-+),\ (--), \\
(+-) &\longrightarrow (++),\ (+-),\ (-+),\ (--), \\
(-+) &\longrightarrow (++),\ (+-),\ (-+),\ (--), \\
(--) &\longrightarrow (++),\ (+-),\ (-+),\ (--).
\end{aligned}
$$

Both possibilities will be illustrated by examples.

6.1 The Statistical Density Operator and the Density Matrix

The statistical density operator $\hat{\varrho}$ of a mixture of N quantum states $|\varphi_n\rangle$ is defined by

$$\hat{\varrho} = \sum_{n=1}^{N} p_n \, |\varphi_n\rangle\langle\varphi_n|. \tag{6.1}$$

The statistical weights $p_n \geq 0$ satisfy the condition

$$\sum_{n=1}^{N} p_n = 1.$$

The fact, that the definition uses an incoherent superposition of states, shows that an ensemble of quantum particles can behave like a classical ensemble.

Matrix elements of the density operator, for example in momentum space (or correspondingly in ordinary space),

$$\langle k'|\hat{\varrho}|k\rangle = \sum_{n} p_n \langle k'|\varphi_n\rangle\langle\varphi_n|k\rangle = \sum_{n} p_n \varphi_n(k')\varphi_n(k)^* \equiv \varrho(k', k) \tag{6.2}$$

form a *statistical matrix* or *density matrix* in the momentum (or space) representation. The diagonal elements are positive definite

$$\varrho(k, k) = \sum_{n} p_n |\varphi_n(k)|^2 \geq 0.$$

For normalised states one obtains the relation[1]

$$\int d^3k \, \varrho(k, k) \equiv \mathrm{tr}[\varrho] = \sum_{n} p_n = 1.$$

An example for the calculation of expectation values of operators, as for example an operator $\hat{A}$, with respect to the ensemble is

$$< \hat{A} > = \mathrm{tr}[[A][\varrho]] = \int d^3k \, \langle k|\hat{A}\hat{\varrho}|k\rangle$$

$$= \int d^3k \int d^3k' \, \langle k|\hat{A}|k'\rangle\langle k'|\hat{\varrho}|k\rangle. \tag{6.3}$$

---

[1] The matrices are written in the form [A]. The trace of a matrix [A] is denoted by tr[A]. If the matrix [A] is explicitly characterised by its elements, the notation is $[A] = (A_{ik})$.

6.1.1 Elastic Scattering of Two Particles

The starting point of the discussion are the scattering amplitudes, which have been introduced in Sect.1.4, either in the μ_s- or the channel spin representation S, M_S. For instance, the scattering amplitude $f^S_{M'_S M_S}(k', k)$ for a channel spin S represents the probability for a transition from a spin state with the *projection* M_S into a state with the projection M'_S. The asymptotic wave function is (compare (1.46))

$$\langle r, 12 | \psi_A \rangle \xrightarrow{r \to \infty} \left\{ e^{ikz} \chi_{M_S} + \sum_{M'_S} f^S_{M'_S M_S}(k', k) \frac{e^{ikr}}{r} \chi_{M'_S} \right\}.$$

The individual scattering amplitudes are interpreted as the elements of a square matrix of dimension $(2S + 1)$

$$[F^S(k', k)] = (f^S_{M'_S M_S}(k', k)) \implies (f^S_{mn}(k', k)),$$

where the second option serves as an abbreviation. If one characterises the spinors for an arbitrary initial as well as final state in addition by a single-column matrix

$$\chi_i = \begin{pmatrix} c_{i_1} \\ c_{i_2} \\ \vdots \\ c_{i(2S+1)} \end{pmatrix} \quad \text{or} \quad \chi_f = \begin{pmatrix} c_{f_1} \\ c_{f_2} \\ \vdots \\ c_{f(2S+1)} \end{pmatrix},$$

one can write the differential cross section (a real quantity) for the transition from an initial state i into a final state f as the square of the absolute value of the corresponding matrix elements in spinor space

$$\frac{d\sigma}{d\Omega} = (\chi_f^\dagger [F^S] \chi_i)^\dagger (\chi_f^\dagger [F^S] \chi_i). \tag{6.4}$$

As an alternative one can use the matrix/spinor form

$$\frac{d\sigma}{d\Omega} = \chi_i^\dagger [F^S]^\dagger \chi_f \chi_f^\dagger [F^S] \chi_i.$$

The term $\chi_f \chi_f^\dagger$ is a $(2S+1)$-dimensional matrix

$$[\chi_f \chi_f^\dagger] = \begin{pmatrix} c_{f_1} \\ c_{f_2} \\ \vdots \\ c_{f_{(2S+1)}} \end{pmatrix} (c_{f_1}^* \; c_{f_2}^* \cdots c_{f_{(2S+1)}}^*)$$

$$= \begin{pmatrix} c_{f_1} c_{f_1}^* & \cdots & c_{f_1} c_{f_{(2S+1)}}^* \\ \vdots & \vdots & \vdots \\ c_{f_{(2S+1)}} c_{f_1}^* & \cdots & c_{f_{(2S+1)}} c_{f_{(2S+1)}}^* \end{pmatrix},$$

(6.5)

which is called the *spin density matrix* $[\varrho_f] = [\chi_f \chi_f^\dagger]$.

An explicit expression for the differential cross section in Eq. (6.4) can be given, if one uses the matrix

$$[B] = [[F^S]^\dagger [\varrho_f][F^S]].$$

The result is

$$\frac{d\sigma}{d\Omega} = \sum_{nn'} c_{i_n}^* B_{nn'} c_{i_{n'}}$$

or alternatively

$$\frac{d\sigma}{d\Omega} = \sum_n \left[\sum_{n'} B_{nn'} \left(c_{i_{n'}} c_{i_n}^* \right) \right].$$

The expression in the round brackets is the spin density matrix for the initial situation. The sum over n' results from the matrix product of $[\varrho_i]$ with the matrix $[B]$, the sum over n from the formation of the trace. The final expression for the differential cross section is the trace of a matrix product

$$\frac{d\sigma}{d\Omega} = \text{tr}[[\varrho_i][F^S]^\dagger [\varrho_f][F^S]].$$

(6.6)

The relation (6.6) is the central point for the specification of various experimental options with polarised beams and targets. Pure states as well as statistical ensembles for the initial and final situation can be considered. This statement will be illustrated by a number of examples in the following.

6.1.2 Spin-1/2 Particles are Scattered by Spin-0 Particles

It is not necessary to state which type of particle constitutes the beam or the target. The roles are interchangeable. In the present simple example all relevant aspects are addressed, so that it is best suited for an introduction of the notation and its handling.

6.1.2.1 The Spin-1/2 Particle Is in a Pure State

The spin-0 particle is not orientated in space, therefore its spin wave function is $\chi(0) = 1$. The density matrix for this collision is determined solely by the spinor of the spin-1/2 particle

$$\chi = \begin{pmatrix} c_1 \\ c_2 \end{pmatrix} \quad \text{with} \quad \chi^\dagger \chi = c_1^* c_1 + c_2^* c_2 = 1.$$

The spin density matrix has the form

$$[\varrho] = [\chi \chi^\dagger] = \begin{pmatrix} c_1 c_1^* & c_1 c_2^* \\ c_2 c_1^* & c_2 c_2^* \end{pmatrix}, \tag{6.7}$$

with $\text{tr}[\varrho] = 1$.

Further properties of the matrix (6.7) are:

- Every (2×2)-matrix can be written as a linear combination of the three Pauli matrices

$$[\sigma]_1 \equiv [\sigma]_x = \begin{pmatrix} 0 & 1 \\ 1 & 0 \end{pmatrix}, \quad [\sigma]_2 \equiv [\sigma]_y = \begin{pmatrix} 0 & -i \\ i & 0 \end{pmatrix}, \quad [\sigma]_3 \equiv [\sigma]_z = \begin{pmatrix} 1 & 0 \\ 0 & -1 \end{pmatrix}$$

and the (2×2) unit matrix

$$[I] = \begin{pmatrix} 1 & 0 \\ 0 & 1 \end{pmatrix}.$$

For the density matrix (6.7) one finds

$$[\varrho] = \frac{1}{2}([I] + \boldsymbol{P} \cdot [\boldsymbol{\sigma}]). \tag{6.8}$$

The three components of the vector $\boldsymbol{P}$ are

$$P_1 = c_1 c_2^* + c_2 c_1^* = 2\text{Re}(c_1 c_2^*),$$
$$P_2 = i(c_1 c_2^* - c_2 c_1^*) = -2\text{Im}(c_1 c_2^*),$$
$$P_3 = c_1 c_1^* - c_2 c_2^*.$$

- The spin density matrix has the eigenvalues 0 and 1, as one has

$$\begin{vmatrix} c_1 c_1^* - r & c_1 c_2^* \\ c_2 c_1^* & c_2 c_2^* - r \end{vmatrix} = r(r-1) \stackrel{!}{=} 0.$$

The corresponding eigenvectors are

$$\varrho_0 = \begin{pmatrix} c_2^* \\ -c_1^* \end{pmatrix} \quad \text{and} \quad \varrho_1 = \begin{pmatrix} c_1 \\ c_2 \end{pmatrix}.$$

They are orthogonal and normalised to 1

$$\varrho_0^\dagger \varrho_0 = \varrho_1^\dagger \varrho_1 = 1,$$
$$\varrho_0^\dagger \varrho_1 = \varrho_1^\dagger \varrho_0 = 0.$$

- The spin density operator is a projection operator with the property

$$\hat{\varrho}^2 = \hat{\varrho}. \tag{6.9}$$

In order to prove this statement, one uses, for example, the definition of the operator in matrix form as in (6.5) with a suitable rearrangement of the factors

$$[\varrho]^2 = [\chi (\chi^\dagger)[\chi) \chi^\dagger] = [\chi \chi^\dagger] = [\varrho].$$

- The calculation of $[\varrho]^2$ with the representation of $[\varrho]$ by the Pauli matrices (6.8) gives

$$[\varrho]^2 = \frac{1}{4} \{[I] + 2(\boldsymbol{P} \cdot [\boldsymbol{\sigma}]) + (\boldsymbol{P} \cdot [\boldsymbol{\sigma}])(\boldsymbol{P} \cdot [\boldsymbol{\sigma}])\}.$$

The Pauli matrices obey anticommutation relations

$$\{[\sigma_k], [\sigma_l]\} = [\sigma_k][\sigma_l] + [\sigma_l][\sigma_k] = \delta_{kl}[I].$$

On the basis of these relations, one finds for the last term in the equation above

$$(\boldsymbol{P} \cdot [\boldsymbol{\sigma}])(\boldsymbol{P} \cdot [\boldsymbol{\sigma}]) = P^2[I].$$

The projection property (6.9) then yields the condition

$$\frac{1}{4}([I] + P^2[I] + 2(\boldsymbol{P} \cdot [\boldsymbol{\sigma}])) = \frac{1}{2}([I] + (\boldsymbol{P} \cdot [\boldsymbol{\sigma}])),$$

from which $P^2 = 1$ follows. The vector $\boldsymbol{P}$ is a vector of length 1.

- For the expectation value of the operator $\hat{\sigma}$ one calculates

$$< \hat{\sigma} > = \text{tr}[[\varrho][\sigma]] = \boldsymbol{P},$$

as the trace of σ vanishes, $\text{tr}[\sigma] = \boldsymbol{0}$ and the trace of a product of two Pauli matrices is proportional to a Kronecker symbol

$$\text{tr}[[\sigma]_k[\sigma]_l] = 2\delta_{kl}.$$

The polarisation vector $\boldsymbol{P}$ is determined by a pure state in this case.

6.1.2.2 Explicit Case Studies

Consider the situation: The target consists of spin-0 particles and the projectiles are spin-1/2 fermions. In order to expose the physical content of an *experiment*, it is useful to parametrise the components of the spinor χ in a suitable way. The choice

$$c_1 = \cos\frac{\theta}{2}, \qquad c_2 = e^{i\delta}\sin\frac{\theta}{2} \tag{6.10}$$

satisfies all requirements. The components of the polarisation vector are then

$$P_1 = \sin\theta\cos\delta, \qquad P_2 = \sin\theta\sin\delta, \qquad P_3 = \cos\theta.$$

The following examples demonstrate the possibilities.

- The polarisation vector has only an x-component. The spin vector points on the *average* in the x-direction. The statement

$$P_1 = 1, \qquad P_2 = P_3 = 0$$

corresponds to

$$\theta = \pi/2, \quad \delta = 0.$$

The polarisation in the x-direction is characterised by a spinor of the form

$$\chi_x = \frac{1}{\sqrt{2}}\begin{pmatrix} 1 \\ 1 \end{pmatrix},$$

and the density matrix is

$$[\varrho]_x = \begin{pmatrix} \frac{1}{2} & \frac{1}{2} \\ \frac{1}{2} & \frac{1}{2} \end{pmatrix}.$$

- For a polarisation in the negative x-direction one finds accordingly

$$P_1 = -1, \quad P_2 = P_3 = 0 \quad \longrightarrow \quad \theta = -\pi/2, \; \delta = 0$$

$$\chi_{(-x)} = \frac{1}{\sqrt{2}} \begin{pmatrix} 1 \\ -1 \end{pmatrix}, \quad [\varrho]_{(-x)} = \begin{pmatrix} \frac{1}{2} & -\frac{1}{2} \\ -\frac{1}{2} & \frac{1}{2} \end{pmatrix}.$$

- A standard situation, which is encountered in experiments, is a polarisation of the beam in the z-direction (projection of the spin onto the beam axis). Here one has

$$P_1 = P_2 = 0, \quad P_3 = 1 \quad \longrightarrow \quad \theta = 0,$$

$$\chi_z = \begin{pmatrix} 1 \\ 0 \end{pmatrix}, \quad [\varrho]_z = \begin{pmatrix} 1 & 0 \\ 0 & 0 \end{pmatrix}.$$

- The polarisation in the negative z-direction is described by

$$P_1 = P_2 = 0, \quad P_3 = -1 \quad \longrightarrow \quad \theta = \pi,$$

$$\chi_{(-z)} = \begin{pmatrix} 0 \\ 1 \end{pmatrix}, \quad [\varrho]_{(-z)} = \begin{pmatrix} 0 & 0 \\ 0 & 1 \end{pmatrix}.$$

The density matrices, which have been calculated, determine the corresponding differential cross sections. As an illustration, one can consider the situation that the fermions are initially prepared with a certain polarisation. In the final state, the spin polarisation in the z-direction is not measured. The differential cross section for this scenario is according to (6.6)

$$\frac{d\sigma}{d\Omega} = \text{tr}[[\varrho_i][F^{1/2}]^\dagger [\varrho_z][F^{1/2}] + \text{tr}[[\varrho_i][F^{1/2}]^\dagger [\varrho_{-z}][F^{1/2}]].$$

As, however, the sum $[\varrho_z] + [\varrho_{(-z)}]$ is equal to the unit matrix, only

$$\frac{d\sigma}{d\Omega} = \text{tr}[[\varrho_i][F^{1/2}]^\dagger [F^{1/2}]]$$

has to be calculated.

The scattering matrices (suppress the upper index)

$$[F^{1/2}] = \begin{pmatrix} f_{1/2,1/2} & f_{1/2,-1/2} \\ f_{-1/2,1/2} & f_{-1/2,-1/2} \end{pmatrix} \equiv \begin{pmatrix} f_{1,1} & f_{1,2} \\ f_{2,1} & f_{2,2} \end{pmatrix}$$

mediate transitions from the second index to the first. In the abbreviated form the replacements $1/2 \rightarrow 1$ and $-1/2 \rightarrow 2$ are used. One still has to calculate the

Fig. 6.1 Spin polarisation
$m_{s_z} = +1/2$

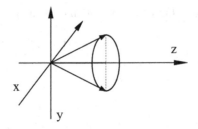

product of the matrices $[F]$ and $[F]^\dagger$ and finds

$$[F]^\dagger[F] = \begin{pmatrix} f_{1,1}^* f_{1,1} + f_{2,1}^* f_{2,1} & f_{1,1}^* f_{1,2} + f_{2,1}^* f_{2,2} \\ f_{1,2}^* f_{1,1} + f_{2,2}^* f_{2,1} & f_{1,2}^* f_{1,2} + f_{2,2}^* f_{2,2} \end{pmatrix},$$

which can be written in abbreviated form as

$$[F]^\dagger[F] = [A] = \begin{pmatrix} A_{1,1} & A_{1,2} \\ A_{2,1} & A_{2,2} \end{pmatrix}.$$

The formulation of the differential cross sections for different scenarios is now no problem.[2] The following examples, in which the positive z-direction was chosen as the beam direction, give an idea of the possibilities.

- For an initial polarisation in the beam direction (that is, the projection of the quantum mechanically fluctuating spin vector onto the beam axis is $+1/2$ for each particle in the beam, Fig. 6.1) one finds for the differential cross section

$$\frac{d\sigma}{d\Omega_z} = tr[[\varrho_z][A]] = tr\left[\begin{pmatrix} 1 & 0 \\ 0 & 0 \end{pmatrix}\begin{pmatrix} A_{1,1} & A_{1,2} \\ A_{2,1} & A_{2,2} \end{pmatrix}\right] = A_{1,1}.$$

- In the case of an initial polarisation in the negative beam direction the cross section is

$$\frac{d\sigma}{d\Omega_{(-z)}} = tr[[\varrho_{-z}][A]] = A_{2,2}.$$

- The beam is initially polarised along the negative x-direction (Fig. 6.2 with, as indicated, a right-handed coordinate system). Here one finds

$$\frac{d\sigma}{d\Omega_{(-x)}} = tr[[\varrho_{(-x)}][A]] = \frac{1}{2}\left(A_{1,1} + A_{2,2} - A_{2,1} - A_{1,2}\right).$$

---

[2] Note also: The scattering amplitudes $f_{m,n}$ are functions of the energy and the scattering angle $f_{m,n} = f_{m,n}(\mathbf{k}, \mathbf{k}')$.

Fig. 6.2 Spin polarisation:
$m_{s_x} = -1/2$

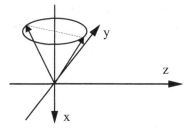

- Similarly, one finds for a polarisation in the direction of the positive x-axis

$$\frac{d\sigma}{d\Omega_x} = \mathrm{tr}[[\varrho_x][A]] = \frac{1}{2}\left(A_{1,1} + A_{2,2} + A_{2,1} + A_{1,2}\right).$$

- One quantity, that is quite sensitive to finer spin effects, is the asymmetry parameter (also called the asymmetry for short), which is defined by

$$A_x = \frac{\left(\dfrac{d\sigma}{d\Omega}\right)_x - \left(\dfrac{d\sigma}{d\Omega}\right)_{(-x)}}{\left(\dfrac{d\sigma}{d\Omega}\right)_x + \left(\dfrac{d\sigma}{d\Omega}\right)_{(-x)}} = \frac{A_{2,1} + A_{1,2}}{A_{1,1} + A_{2,2}}$$

or by

$$A_z = \frac{\left(\dfrac{d\sigma}{d\Omega}\right)_z - \left(\dfrac{d\sigma}{d\Omega}\right)_{(-z)}}{\left(\dfrac{d\sigma}{d\Omega}\right)_z + \left(\dfrac{d\sigma}{d\Omega}\right)_{(-z)}} = \frac{A_{1,1} - A_{2,2}}{A_{1,1} + A_{2,2}}.$$

In the actual experiment, the *polariser*, which adjusts the orientation of the spin, is rotated by 180° and the corresponding particle numbers are measured in the detector for the two settings

$$A_{x,z}(\text{experiment}) = \frac{N_+ - N_-}{N_+ + N_-}.$$

6.1.2.3 Statistical Mixture for the Spin-1/2 Particles
If there is a statistical mixture, the statistical density operator is given by a linear combination of density operators with pure states

$$\hat{\varrho}_{\text{ens}} \equiv \hat{\varrho} = \sum_n P_n \hat{\chi}_n \hat{\chi}_n^\dagger = \sum_n P_n \hat{\varrho}_n. \tag{6.11}$$

An example is the density operator for an unpolarised beam – a mixture of an equal number of oppositely polarised particles (either in x- or in z-direction)

$$\hat{\varrho}_{\text{unpol}} = \frac{1}{2}\hat{\varrho}_x + \frac{1}{2}\hat{\varrho}_{(-x)} = \frac{1}{2}\hat{\varrho}_z + \frac{1}{2}\hat{\varrho}_{(-z)} = \frac{1}{2}\begin{pmatrix} 1 & 0 \\ 0 & 1 \end{pmatrix}.$$

The states, which contribute to an ensemble (6.11) do not necessarily have to be orthogonal. However, even if they are, the statement is

$$\hat{\varrho}^2 \neq \hat{\varrho}.$$

For example, a mixture of two spin states is described by

$$\hat{\varrho}^2 = (p_1\hat{\chi}_1\hat{\chi}_1^\dagger + p_2\hat{\chi}_2\hat{\chi}_2^\dagger)(p_1\hat{\chi}_1\hat{\chi}_1^\dagger + p_2\hat{\chi}_2\hat{\chi}_2^\dagger).$$

If the two spinors are orthogonal, one finds

$$\hat{\varrho}^2 = p_1^2\hat{\chi}_1\hat{\chi}_1^\dagger + p_2^2\hat{\chi}_2\hat{\chi}_2^\dagger \neq \hat{\varrho},$$

that is

$$\hat{\varrho}^2_{\text{unpol}} = \frac{1}{2}\hat{\varrho}_{\text{unpol}}$$

in the scenario indicated above.

Some explicit examples for the calculation of differential cross sections are:

• For an unpolarised beam (e.g. with respect to the z-direction) the following result is found

$$\frac{d\sigma}{d\Omega_{\text{unpol}}} = \frac{1}{2}\{\text{tr}[[\varrho_z][A]] + \text{tr}[[\varrho_{-z}][A]]\}$$

$$= \frac{1}{2}\text{tr}\left\{\begin{pmatrix} 1 & 0 \\ 0 & 1 \end{pmatrix}\begin{pmatrix} A_{1,1} & A_{1,2} \\ A_{2,1} & A_{2,2} \end{pmatrix}\right\} = \frac{1}{2}(A_{1,1} + A_{2,2})$$

$$= \frac{1}{2}\{f_{1,1}^* f_{1,1} + f_{2,1}^* f_{2,1} + f_{1,2}^* f_{1,2} + f_{2,2}^* f_{2,2}\}$$

$$= \frac{1}{2}\sum_{\mu_i,\mu_f} f_{\mu_i,\mu_f}^* f_{\mu_i,\mu_f}.$$

This result has already been obtained by elementary means in Sect. 1.4.

- The beam is polarised with 30 % in the $+x$- (or $-x$-) direction and 35 % each in the $\pm z$ directions. The density operator is in these cases

$$[\varrho_{x,z}] = \frac{3}{10} \begin{pmatrix} \frac{1}{2} & \frac{1}{2} \\ \frac{1}{2} & \frac{1}{2} \end{pmatrix} + \frac{35}{100} \begin{pmatrix} 1 & 0 \\ 0 & 0 \end{pmatrix} + \frac{35}{100} \begin{pmatrix} 0 & 0 \\ 0 & 1 \end{pmatrix} = \frac{1}{2} \begin{pmatrix} 1 & \frac{3}{10} \\ \frac{3}{10} & 1 \end{pmatrix},$$

$$[\varrho_{-x,-z}] = \frac{3}{10} \begin{pmatrix} \frac{1}{2} & -\frac{1}{2} \\ -\frac{1}{2} & \frac{1}{2} \end{pmatrix} + \frac{35}{100} \begin{pmatrix} 1 & 0 \\ 0 & 1 \end{pmatrix} = \frac{1}{2} \begin{pmatrix} 1 & -\frac{3}{10} \\ -\frac{3}{10} & 1 \end{pmatrix}.$$

For the cross section and for the asymmetry parameter one obtains in this situation

$$\frac{d\sigma}{d\Omega}_{(\pm x)} = \text{tr}[[\varrho_{(\pm x)}][A]] = \frac{1}{2}\left(A_{1,1} + A_{2,2} \pm \frac{3}{10}\{A_{2,1} + A_{1,2}\}\right)$$

and

$$A_x = \frac{\left(\dfrac{d\sigma}{d\Omega}\right)_+ - \left(\dfrac{d\sigma}{d\Omega}\right)_-}{\left(\dfrac{d\sigma}{d\Omega}\right)_+ + \left(\dfrac{d\sigma}{d\Omega}\right)_-} = \frac{3}{10} \frac{(A_{2,1} + A_{1,2})}{(A_{1,1} + A_{2,2})}.$$

These results can be used to extract statements about the composition of the two beams in comparison with results of experiments with pure beams.

6.1.3 Spin-1/2 Particles are Scattered by Spin-1/2 Particles

The μ_s-representation with four spinor combinations will be used in this instance. The form of the spinors of the two-fermion system is

$$\chi = \begin{pmatrix} c_1(1)c_1(2) \\ c_1(1)c_2(2) \\ c_2(1)c_1(2) \\ c_2(1)c_2(2) \end{pmatrix} \equiv \begin{pmatrix} c_1 \\ c_2 \\ c_3 \\ c_4 \end{pmatrix}.$$

The index denotes the state of the spin, the argument denotes the particle. The spin density matrix is a (4×4)-matrix

$$[\varrho] = [\chi \chi^\dagger] = \begin{pmatrix} c_1 c_1^* & c_1 c_2^* & c_1 c_3^* & c_1 c_4^* \\ c_2 c_1^* & c_2 c_2^* & c_2 c_3^* & c_2 c_4^* \\ c_3 c_1^* & \cdots & & \cdots \\ c_4 c_1^* & \cdots & & c_4 c_4^* \end{pmatrix}.$$

The cross sections are calculated, as shown in (6.6), in the following fashion

$$\frac{d\sigma}{d\Omega} = \text{tr}[[\varrho_i][F^S]^\dagger[\varrho_f][F^S]],$$

or according to

$$\frac{d\sigma}{d\Omega} = \text{tr}[[\varrho_i][F^S]^\dagger[[F^S]] \equiv \text{tr}[[\varrho_i][A]],$$

if the final situation is not analysed with respect to spin. The matrix $[A]$ is also a (4×4) matrix, corresponding to the 16 possible transitions in this collision system.

The results for four examples, which the reader may try as an exercise, are given below. In the first two examples the initial states are pure states, in the last two the initial states are ensembles.

- Both particles are in a state with the projection $+1/2$ in the z-direction. This is expressed by

$$c_1(1) = 1, \ c_2(1) = 0, \ c_1(2) = 1, \ c_2(2) = 0.$$

The spin density matrix is therefore

$$[\varrho]_{++} = \begin{pmatrix} 1 & 0 & 0 & 0 \\ 0 & 0 & 0 & 0 \\ 0 & 0 & 0 & 0 \\ 0 & 0 & 0 & 0 \end{pmatrix}.$$

The cross section is

$$\frac{d\sigma}{d\Omega} = A_{11} = f_{11}^* f_{11} + f_{21}^* f_{21} + f_{31}^* f_{31} + f_{41}^* f_{41}.$$

The particles make a transition, with a probability depending on the energy and the scattering angle, from the state $(++)$ into the states $(++)$, $(+-)$, $(-+)$ and $(--)$.

- Particle 1 is in the state $+1/2$, particle 2 in the state $-1/2$ with respect to the z-direction. Here the initial situation is

$$c_1(1) = 1, \ c_2(1) = 0, \ c_1(2) = 0, \ c_2(2) = 1.$$

The density matrix

$$[\varrho]_{+-} = \begin{pmatrix} 0\ 0\ 0\ 0 \\ 0\ 1\ 0\ 0 \\ 0\ 0\ 0\ 0 \\ 0\ 0\ 0\ 0 \end{pmatrix}$$

leads to the cross section

$$\frac{d\sigma}{d\Omega} = A_{22} = f_{12}^* f_{12} + f_{22}^* f_{22} + f_{32}^* f_{32} + f_{42}^* f_{42}.$$

Four possible final states can be reached, starting from the initial state $(+-)$.
- An unpolarised beam scatters from an unpolarised target. The starting point is in this case

$$[\varrho]_{\mathrm{unpol}} = \frac{1}{4}\left([\varrho]_{++} + [\varrho]_{+-} + [\varrho]_{-+} + [\varrho]_{--} \right) = \frac{1}{4}[I],$$

so that one obtains

$$\frac{d\sigma}{d\Omega} = \frac{1}{4}\left(A_{11} + A_{22} + A_{33} + A_{44} \right).$$

The system goes from all possible initial states to all possible final states. The weight for each of the initial states is $1/4$. As no measurement of the final spin configuration is performed, the final density matrix is $[\varrho]_f = [I]$.
- The beam is polarised in the positive x-direction to P % and in the positive z-direction to $(100 - P)$ %, the target is polarised in the positive z-direction. For the pure states one has in this case

$$\chi_A = \begin{pmatrix} \frac{1}{\sqrt{2}} \\ 0 \\ \frac{1}{\sqrt{2}} \\ 0 \end{pmatrix}$$

and

$$X_B = \begin{pmatrix} 1 \\ 0 \\ 0 \\ 0 \end{pmatrix}.$$

From this follows

$$[\varrho]_A = \begin{pmatrix} \frac{1}{2} & 0 & \frac{1}{2} & 0 \\ 0 & 0 & 0 & 0 \\ \frac{1}{2} & 0 & \frac{1}{2} & 0 \\ 0 & 0 & 0 & 0 \end{pmatrix} \quad \text{and} \quad [\varrho]_B = \begin{pmatrix} 1 & 0 & 0 & 0 \\ 0 & 0 & 0 & 0 \\ 0 & 0 & 0 & 0 \\ 0 & 0 & 0 & 0 \end{pmatrix}.$$

The density matrix for this example is obtained from the density operators of the pure states, each weighted by

$$p_A = \frac{P}{100} = p \quad \text{or} \quad p_B = 1 - p$$

so that the result for the density matrix is

$$[\varrho] = p[\varrho]_A + (1 - p)[\varrho]_B = \begin{pmatrix} 1 - \frac{p}{2} & 0 & \frac{p}{2} & 0 \\ 0 & 0 & 0 & 0 \\ \frac{p}{2} & 0 & \frac{p}{2} & 0 \\ 0 & 0 & 0 & 0 \end{pmatrix}.$$

The result for the cross section is therefore

$$\frac{d\sigma}{d\Omega} = (1 - \frac{p}{2})A_{11} + \frac{p}{2}(A_{13} + A_{31} + A_{33}).$$

The formalism described is not only used in single scattering experiments, but also in double scattering experiments. In this case, one typically first uses an unpolarised beam which is scattered by a target and thereby gains a polarisation $P_1(\theta_1)$ in the θ_1 direction. To determine this polarisation, which can give information about the interaction between target and projectile, one scatters the emerging beam from a second target of the same kind and measures the asymmetry at the angles $\pm\theta_2$. The asymmetry is then

$$A = P_1(\theta_1)P_1(\theta_2),$$

so that one can determine the function $P_1(\theta)$ by variation of the two angles.

Remarks on Multichannel Problems

7

The discussion of multichannel collision problems should really occupy more space than the discussion of elastic scattering. This relatively short chapter is only a first introduction to this subject. The key word is the word *channel* with specifications as elastic channel, particle transfer channel, ionisation channel and many others. The quantity that is quite helpful in sorting the possible reaction channels is the energy.

A system with three different particles will be used for the sake of simplicity for the discussion of key concepts instead of a realistic collision system. In addition to the kinetic energy of the particles, there is an interaction between each pair of particles. The particles are chosen to be different in order to avoid a discussion of the exchange symmetry with antisymmetrisation or symmetrisation. After sorting the reaction channels, the Møller operators for possible entrance and exit channels are introduced. The Møller operators for the entrance channels $\hat{\Omega}_+^\alpha$ describe the time evolution in the channel α from the initial time $t = -\infty$ up to the time $t = 0$, the operators $\hat{\Omega}_-^\beta$ for the exit channels from the time $t = \infty$ back to the time $t = 0$. These operators allow the construction of the formal theory, but the application of the theory can not be implemented in a direct fashion. This is demonstrated by an outline of the Faddeev three-body problem.

As in the case of one entrance and one exit channel, discussed in Chap. 3, one can construct the S-matrix operators $\hat{S}_{\alpha \to \beta}$ for the transition from the initial channel α into final channels β with the aid of Møller operators. After the introduction of the generalisation of the channel-to-channel S-matrix elements, the T-matrix elements, which are matrix elements of an operator with respect to plane wave states in the two channels involved $\langle \alpha K' | \hat{T} | \beta K \rangle$, can be considered. The partition of the reaction process into two steps, allows a distinction between two types of T-matrix elements:

- The *entrance channel form* (prior form), in which the T-matrix element involves the interaction, that is not part of the asymptotic Hamiltonian of the entrance channel.

© Springer-Verlag GmbH Germany, part of Springer Nature 2022
R. M. Dreizler et al., *Quantum Collision Theory of Nonrelativistic Particles*,
https://doi.org/10.1007/978-3-662-65591-7_7

- The *exit channel form* (post form), in which the T-matrix element involves the interaction, that is not present in the asymptotic Hamiltonian of the exit channel under consideration.

In consequence of this distinction, one can also discuss two variants of Lippmann-Schwinger equations, which can be useful for the formulation of approximations. Exact solutions of the two types of equations would lead to the same result, but approximations in the post and the prior form usually give different results.

The three-body collision problem formulated by Faddeev is based on a variant of the model system indicated above. One first finds (most directly by a representation of the relevant equations in terms of Feynman diagrams), that the standard form of the integral equations leads to singular solutions. It is, however, possible to modify the structure of the equations, in a way that singular solutions do not occur. The method used for this purpose is an expansion in terms of T-matrix elements of two-particle subsystems instead of an expansion in terms of the matrix elements of the two-particle interactions involved. In addition to the analysis of this explicit three-body problem, a brief excursion into nuclear physics is used to illustrate the application of the multichannel formulation by a discussion of direct nuclear reactions based on an abbreviated version of the theory of (d, p) processes.

7.1   Channels and Channel Møller Operators

A collision system with several final channels from the field of atomic physics is the scattering of an electron by a hydrogen atom. Possible processes are

$$
\begin{aligned}
e + H &\longrightarrow e + H & \text{elastic scattering}\\
&\longrightarrow e + H^* & \text{excitation}\\
&\longrightarrow e + e + p & \text{ionisation}\\
&\longrightarrow \dots &
\end{aligned}
$$

As the energy of the projectile electron increases, an increasing number of bound states of the atom are excited. Ionisation occurs in addition to the individual excitation channels. The ionisation of an atomic electron by a projectile electron is referred to as an (e, 2e) process. It is characterised by the exchange symmetry of the two electrons and the long-range Coulomb interaction.

The following modelistic three-body problem is used here to illustrate the multichannel situation instead of a realistic system, in which the reactions are initiated by an electron impinging on a hydrogen atom. This system consists of three *distinguishable* particles a, b, c. An implementation of exchange symmetry is not necessary in this case. The Hamiltonian of the system has the form

$$
\hat{H} = -\frac{\hbar^2 \Delta_a}{2m_a} - \frac{\hbar^2 \Delta_b}{2m_b} - \frac{\hbar^2 \Delta_c}{2m_c} + V_{ab}(\boldsymbol{r}_{ab}) + V_{ac}(\boldsymbol{r}_{ac}) + V_{bc}(\boldsymbol{r}_{bc}). \tag{7.1}
$$

Besides the kinetic energies of the particles, only interactions between pairs of particles are present, assuming that, unlike the Coulomb interaction in real problems, their range is short. The interactions are supposed to be such that the subsystems

$$(bc), \ (ac), \ (abc)$$

have a finite number of bound states. No bound states exist if a pair of particles is broken up $(ab) = a + b$.

If the initial channel is the collision of the particle a with the bound system (bc) in the ground state, the processes listed in the table below are possible. The initial kinetic energy of particle a at a sufficiently large distance from the bound system (bc) is E_a. It is measured with respect to the ground-state energy $E_0(bc) = 0$. The various channels listed in the table are

entrance	exit	channel number
$a + (bc) \longrightarrow$	$a + b + c$ (break up)	0
	$a + (bc)$ (elastic scattering)	1
	$a + (bc)^*$ (excitation)	2
	$b + (ac)$ (transfer)	3
	$b + (ac)^*$ (transfer excitation)	4
	(abc) (capture)	5
	$(abc)^*$ (capture and excitation)	6

It is useful to enumerate the exit channels consecutively, although

- the order is actually arbitrary,
- the different modes of excitation are not differentiated. This could be included by means of a more detailed sequence of numbers 2.1, 2.2, ..., 4.1, ... etc.

Different threshold energies exist for the distinct channels. For example, elastic scattering of a is possible for all energies with $E_a > 0$. In order to excite the system (bc) to a state with the excitation energy $\Delta E_{(bc)}$ above its ground state, the projectile must have at least the energy $E_a \geq \Delta E_{(bc)}$.

For the elastic channel $a + (bc)$ the Hamiltonian acting in the asymptotic region ($t \to \pm\infty$ in the time-dependent formulation) is

$$\hat{H}^{(1)} = \hat{T}^a + \hat{T}^b + \hat{T}^c + \hat{V}_{(bc)}.$$

A solution of the Schrödinger equation for this Hamiltonian with appropriate boundary conditions - the channel with index 1 - describes the motion of particle a in the direction of or away from the bound system (bc) in the ground state. The stationary wave function of the asymptotic state in the range $|\boldsymbol{r}_a - \boldsymbol{R}_{(bc)}| \to \infty$ has the form

$$\langle r_a r_b r_c | 1; k_a, k_{(bc)} \rangle \longrightarrow \phi_{k_{(bc)},0}(r_{(bc)}) \psi_{k_a - k_{(bc)}}(r_a - R_{(bc)}).$$

The function ϕ_0 represents the ground state of the system (bc), whose motion is characterised by the wave number $k_{(bc)}$. The function ψ describes the motion of a with respect to the centre of mass of (bc).

The channel 2 is characterised in the asymptotic domain by the same Hamiltonian

$$\hat{H}^{(2)} = \hat{H}^{(1)}.$$

The wave function of the corresponding asymptotic states with excitation of the target is

$$\langle r_a r_b r_c | 2.i; k_a, k_{(bc)} \rangle \longrightarrow \phi_{k_{(bc)},i}(r_{(bc)}) \psi_{k_a - k_{(bc)}}(r_a - R_{(bc)})$$

$$i = 1, 2, \ldots \quad .$$

The asymptotic Hamiltonian for the transfer channels 3 and 4 is

$$\hat{H}^{(3)} \equiv \hat{H}^{(4)} = \hat{T}^a + \hat{T}^b + \hat{T}^c + \hat{V}_{(ac)}.$$

The corresponding asymptotic states (with $|r_b - R_{(ac)}| \to \infty$) for the various possible channels of 4 are for example

$$\langle r_a r_b r_c | 4.i; k_b, k_{(ac)} \rangle \longrightarrow \phi_{k_{(ac)},i}(r_{(ac)}) \psi_{k_b - k_{(ac)}}(r_b - R_{(ac)})$$

$$i = 1, 2, \ldots \quad .$$

A useful, general form for the *asymptotic states in a channel γ* is $|\gamma K\rangle$ with

$$\hat{H}^{(\gamma)} |\gamma K\rangle = E(\gamma K) |\gamma K\rangle.$$

The quantity K characterises all wave numbers needed to identify the channel. The energy $E(\gamma, K)$ is the channel energy. Instead of a complete characterisation of the energy and the states, one may use the abbreviation

$$\hat{H}^{(\gamma)} |\gamma\rangle = E(\gamma) |\gamma\rangle.$$

For each possible *final channel β* one can define Møller operators $\hat{\Omega}_-^{(\beta)}$. These *channel Møller operators* with

$$|\beta, -\rangle = \hat{\Omega}_-^{(\beta)} |\beta\rangle = \lim_{t \to +\infty} e^{i\frac{\hat{H}t}{\hbar}} e^{-i\frac{\hat{H}^{(\beta)}t}{\hbar}} |\beta\rangle \tag{7.2}$$

indicate the fact, that, for this channel, only the *channel Hamiltonian* $\hat{H}^{(\beta)}$ is effective at late times. They correspond to an extension of the definitions in Sect. 3.1. For the *entrance channel* α one has

$$|\alpha, +\rangle = \hat{\Omega}_+^{(\alpha)}|\alpha\rangle = \lim_{t \to -\infty} e^{\frac{i\hat{H}t}{\hbar}} e^{-\frac{i\hat{H}^{(\alpha)}t}{\hbar}} |\alpha\rangle. \tag{7.3}$$

As in the case of a single exit channel, the action of the Møller operators on the asymptotic states describes the exact scattering states. The sign $\pm$ indicates whether an *incoming* or whether an *outgoing* state is addressed. In this fashion, the state $|3, K, +\rangle \equiv |3, +\rangle$ represents a state of an entrance channel, in which particle b collides with the system (ac) in the ground state and evolves from this state at the time $t = -\infty$ to the time $t = 0$ under the influence of the total Hamiltonian.

7.2 The Multichannel S-Matrix

The channel Møller operators in (7.2) and (7.3) allow a generalisation of the S-matrix for elastic scattering, which has been discussed in Sect. 3.2 and 3.3. The operator

$$\hat{S}_{\alpha \to \beta} \equiv \hat{S}_{\alpha\beta} = \left(\hat{\Omega}_-^{(\beta)}\right)^\dagger \hat{\Omega}_+^{(\alpha)} \tag{7.4}$$

characterises a transition from an entrance channel α into a final channel β. In extension of the situation in Chap. 3 a product of different channel Møller operators has to be used here.

The matrix elements of these operators are a measure of the probability for a transition from the channel α into the channel β

$$W_{\alpha \to \beta} = |\langle \beta K' - |\alpha K+\rangle|^2 = |\langle \beta K'|\hat{S}_{\alpha\beta}|\alpha K\rangle|^2. \tag{7.5}$$

The matrix elements represent, in the sense of the discussion in Chap. 3, a projection. The state $|\alpha K\rangle$ evolves from the time $t = -\infty$ up to the time $t = 0$. At the time $t = 0$ it is projected onto a state which evolves from $|\beta K\rangle$ at the time $t = +\infty$ backwards in time until $t = 0$.

In the same way as in the proof of the relation (3.23), one shows that the statement

$$\hat{H}\hat{\Omega}_\pm^{(\alpha)} = \hat{\Omega}_\pm^{(\alpha)}\hat{H}^{(\alpha)} \tag{7.6}$$

is valid for the channel Møller operators. After some steps (details in Sect. 7.7.1) one arrives at the relation

$$\left[E(\alpha, K) - E(\beta, K')\right]\langle \beta K'|\hat{S}_{\alpha\beta}|\alpha K\rangle = 0$$

for the S-matrix elements. These elements are different from zero, if and only if the asymptotic initial energy equals the final energy. Another consequence of the relation (7.6) is the statement

$$\hat{H}|\alpha K\pm\rangle \overset{!}{=} \bar{E}(\alpha K)|\alpha K\pm\rangle = \hat{H}\hat{\Omega}_{\pm}^{(\alpha)}|\alpha K\rangle = \hat{\Omega}_{\pm}^{(\alpha)}\hat{H}|\alpha K\rangle$$

$$= E(\alpha K)\hat{\Omega}_{\pm}^{(\alpha)}|\alpha K\rangle = E(\alpha K)|\alpha K\pm\rangle.$$

The total energy $\bar{E}$ in a channel is identical with the *channel energy* E, which is defined via the entrance channel. The energy is then distributed between the binding energy and the kinetic energies of the fragments in each exit channel. This distribution is the reason for the presence of energy thresholds, which has been indicated above. For example, for a three-particle system with

$$E_{(bc)} < E_{(bc)}^* < \ldots < E_{(ac)} < E_{(ac)}^* < \ldots$$

and an energy value $E_\alpha = E_{(bc)} + E_{\text{kin},a}$ in the entrance [1] channel α one finds

- only elastic scattering is possible if $E_{(bc)} \leq E_\alpha \leq E_{(bc)}^*$.
- For a higher energy of the projectile a with $E_{(bc)}^* \leq E_\alpha \leq E_{(ac)}$ excitation of the system (bc) is possible (depending on the impact energy) besides elastic scattering.
- For $E_\alpha \geq E_{(ac)}$ excitation of the system (bc) and particle transfer are possible besides elastic scattering, each with a certain probability.

7.3   The Multichannel Lippmann-Schwinger Equation

The time development operator, which is the basis for the definition of the channel Møller operators, is

$$\hat{O}^{(\gamma)}(t) = e^{i\hat{H}t/\hbar}e^{-i\hat{H}^{(\gamma)}t/\hbar}.$$

Differentiation with respect to time gives

$$-i\hbar\,\partial_t\,\hat{O}^{(\gamma)}(t) = e^{i\hat{H}t/\hbar}\left[\hat{H} - \hat{H}^{(\gamma)}\right]e^{-i\hat{H}^{(\gamma)}t/\hbar} = e^{i\hat{H}t/\hbar}\,\hat{V}^{(\gamma)}\,e^{-i\hat{H}^{(\gamma)}t/\hbar}.$$

The potential, which could be named *channel potential* or in this case explicitly *potential in channel* γ

$$\hat{V}^{(\gamma)} = \hat{H} - \hat{H}^{(\gamma)} \tag{7.7}$$

---

[1] $E_{(bc)} = 0$ if the bound system (bc) is in the ground state at the initial time.

is that part of the Hamiltonian, which is *not* effective in the channel γ (short range assumed). Integration of the expression for the derivative of $\hat{O}^{(\gamma)}(t)$ yields, with $\hat{O}^{(\gamma)}(0) = \hat{1}$,

$$\hat{O}^{(\gamma)}(t) = \hat{1} + \frac{i}{\hbar} \int_0^t dt'\, e^{i\hat{H}t'/\hbar}\, \hat{V}^{(\gamma)}\, e^{-i\hat{H}^{(\gamma)}t'/\hbar}.$$

The limiting values for $t \to \pm\infty$ allow a useful form of the Møller operators

$$\hat{O}_{\pm}^{(\gamma)}(t = \mp\infty) \to \hat{\Omega}_{\pm}^{(\gamma)},$$

provided the operators act on wave packets. If they act on plane wave states (as stipulated above), adiabatic switching with the convergence factor

$$\lim_{\epsilon \to 0} \int \ldots e^{\pm\epsilon t/\hbar} \ldots$$

is required. If this precaution is applied, one obtains the relations

$$|\gamma, \pm\rangle = \hat{\Omega}_{\pm}^{(\gamma)} |\gamma\rangle$$

$$= |\gamma\rangle + \frac{i}{\hbar} \lim_{\epsilon \to 0} \int_0^{\mp\infty} dt\, e^{\pm\epsilon t/\hbar} e^{i\hat{H}t/\hbar}\, \hat{V}^{(\gamma)}\, e^{-i\hat{H}^{(\gamma)}t/\hbar} |\gamma\rangle \qquad (7.8)$$

$$= |\gamma\rangle + \frac{i}{\hbar} \lim_{\epsilon \to 0} \int_0^{\mp\infty} dt\, e^{[-i(E(\gamma)\pm i\epsilon - \hat{H})t/\hbar]}\, \hat{V}^{(\gamma)} |\gamma\rangle.$$

The expression (7.8) contains the Fourier representation of the *exact Green's functions in the channel* γ

$$\hat{G}^{(\pm)}(E(\gamma)) = \frac{1}{(E(\gamma) - \hat{H} \pm i\epsilon)}. \qquad (7.9)$$

This definition permits the extraction of the formal *channel Lippmann-Schwinger equation* from (7.8), which reads

$$|\gamma, \pm\rangle = |\gamma\rangle + \hat{G}^{(\pm)}(E(\gamma))\, \hat{V}^{(\gamma)} |\gamma\rangle. \qquad (7.10)$$

Equation (7.10) can be used for a reformulation of the S-matrix elements defined in (7.5)

$$\langle \beta | \hat{S}_{\alpha\beta} | \alpha \rangle = \langle \beta, -|\alpha, +\rangle \qquad (7.11)$$

by first establishing a useful relation for the Lippmann-Schwinger equations of a channel α in the form

$$|\alpha, +\rangle = |\alpha, -\rangle + \left[\hat{G}^{(+)}(E(\alpha)) - \hat{G}^{(-)}(E(\alpha)) \right] \hat{V}^{(\alpha)}|\alpha\rangle.$$

If one then inserts this relation on the right-hand side of (7.11) and assumes that the (exact) scattering states are orthogonal for the same boundary conditions[2], one finds

$$\langle \beta K'|\hat{S}_{\alpha\beta}|\alpha K\rangle = \delta_{\alpha\beta} \, \delta(K - K')$$
$$+ \langle \beta K' - | \left[\hat{G}^{(+)}(E(\alpha)) - \hat{G}^{(-)}(E(\alpha)) \right] \hat{V}^{(\alpha)}|\alpha K\rangle.$$

The δ-function with the difference of the K-wave vectors stands for the product of the δ-functions of all relevant wave numbers in the channels α and β. As the state $|\beta K'-\rangle$ is an eigenstate of the Hamiltonian $\hat{H}$ with the energy $E(\beta, K')$, the Dirac identity (2.25) leads to the explicit relation

$$\langle \beta K'|\hat{S}_{\alpha\beta}|\alpha K\rangle = \delta_{\alpha\beta}\delta(K - K')$$
$$- 2\pi i \, \delta(E(\beta, K') - E(\alpha, K))\langle \beta K' - |\hat{V}^{(\alpha)}|\alpha K\rangle. \tag{7.12}$$

In analogy to the definition in Sect. 2.2 one recognises the *on-shell* (more precisely on-energy-shell) T-matrix element of the multichannel problem

$$\langle \beta K'|\hat{T}_{\alpha\beta}|\alpha K\rangle = \langle \beta K' - |\hat{V}^{(\alpha)}|\alpha K\rangle. \tag{7.13}$$

Only on-shell T-matrix elements are needed for the calculation of S-matrix elements.

7.4   The Multichannel T-Matrix

The potential $\hat{V}^{(\alpha)}$ in Eq. (7.13) is the potential that does not occur in the asymptotic Hamiltonian $\hat{H}^{(\alpha)}$. One refers to the form of the corresponding T-matrix element

$$\langle \beta K'|\hat{T}_{\alpha\beta}^{(pr)}|\alpha K\rangle = \langle \beta K' - |\hat{V}^{(\alpha)}|\alpha K\rangle$$

as the

- entrance channel form or prior form.

---

[2] Is this statement correct? Verify.

A reformulation of the S-matrix element with the representation of the bra-vector $\langle \beta K' - |$ in terms of Green's functions can be carried out in the same fashion. One obtains in this instance

$$\langle \beta K' | \hat{T}^{(po)}_{\alpha\beta} | \alpha K \rangle = \langle \beta K' | \hat{V}^{(\beta)} | \alpha K + \rangle, \tag{7.14}$$

which is referred to as the

- exit channel form or post form.

As will be shown below, both forms give the same result for on-shell T-matrix elements when the problem is solved exactly.

The Lippmann-Schwinger equation (7.10) with the exact Green's function is useful for general investigations, but not for the formulation of practical approximations. For this purpose, one uses channel Green's functions, which, for a channel γ, are formally defined by

$$\hat{G}^{(\pm)}_{\gamma}(E(\gamma)) = (E(\gamma) - \hat{H}^{(\gamma)} \pm i\epsilon)^{-1}.$$

The operator identity used in Sect. 2.1.1

$$\hat{A}^{-1} = \hat{B}^{-1} + \hat{B}^{-1}(\hat{B} - \hat{A})\hat{A}^{-1}$$

gives, for $\hat{A} = (E - \hat{H} \pm i\epsilon)$ and $\hat{B} = (E - \hat{H}^{(\gamma)} \pm i\epsilon)$, a relation between the exact Green's function and each of the channel-Green's functions (both with the energy E)

$$\hat{G}^{(\pm)}(E) = \hat{G}^{(\pm)}_{\gamma}(E) + \hat{G}^{(\pm)}_{\gamma}(E)\hat{V}^{(\gamma)}\hat{G}^{(\pm)}(E). \tag{7.15}$$

If one exchanges the roles of the operators $\hat{A}$ and $\hat{B}$, one finds the alternative integral equation

$$\hat{G}^{(\pm)}(E) = \hat{G}^{(\pm)}_{\gamma}(E) + \hat{G}^{(\pm)}(E)\hat{V}^{(\gamma)}\hat{G}^{(\pm)}_{\gamma}(E). \tag{7.16}$$

One can show in the next step with (7.15) the validity of the relation

$$\hat{G}^{(\pm)}_{\gamma}(E(\gamma))\hat{V}^{(\gamma)}|\gamma, \pm\rangle = \hat{G}^{(\pm)}(E(\gamma))\hat{V}^{(\gamma)}|\gamma\rangle$$

and obtain instead of (7.10) the integral equation

$$|\gamma, \pm\rangle = |\gamma\rangle + \hat{G}^{(\pm)}_{\gamma}(E(\gamma)) \, \hat{V}^{(\gamma)} |\gamma, \pm\rangle, \tag{7.17}$$

which is called the channel Lippmann-Schwinger equation for the channel γ. These equations give the impression that the N-channel problem can be reduced to N single-channel problems. The impression is deceptive, as the state $|\gamma\pm\rangle$ which evolves from the initial state $|\gamma\rangle$ is not an exact solution. There are two reasons why the solution of the N-channel problem is significantly more difficult than the solution of N single-channel problems:

- The Green's functions that appear in Eq. (7.17) can not, in general, be computed exactly because they describe the motion in a potential. For example, the Green's function for channel 1 of the example in Sect. 7.1 is

$$\hat{G}_1^{(\pm)}(E) = (E \pm i\epsilon - \hat{T}_a - \hat{T}_b - \hat{T}_c - \hat{V}_{(bc)})^{-1}.$$

- When one attempts to solve the integral equation (7.10) or the set of integral equations (7.17) singularities can occur. The reason for this difficulty and its resolution is addressed in Sect. 7.5.

The Lippmann-Schwinger equation (7.10) can be used to obtain explicit equations for the T-matrix operators. If one starts with the prior form

$$\langle \beta | \hat{T}_{\alpha\beta}^{(pr)} | \alpha \rangle = \langle \beta, - | \hat{V}^{(\alpha)} | \alpha \rangle \qquad (7.18)$$

and replaces the bra-vector on the right-hand side by the Lippmann-Schwinger equation (7.10)

$$\langle \beta, - | = \langle \beta | + \langle \beta | \hat{V}^{(\beta)} \hat{G}^{(+)},$$

one finds (use $E(\beta) = E(\alpha) = E$)

$$\langle \beta | \hat{T}_{\alpha\beta}^{(pr)} | \alpha \rangle = \langle \beta | \left\{ \hat{V}^{(\alpha)} + \hat{V}^{(\beta)} \hat{G}^{(+)}(E) \hat{V}^{(\alpha)} \right\} | \alpha \rangle$$

or the corresponding operator equation

$$\hat{T}_{\alpha\beta}^{(pr)} = \hat{V}^{(\alpha)} + \hat{V}^{(\beta)} \hat{G}^{(+)}(E) \hat{V}^{(\alpha)}. \qquad (7.19)$$

If one starts from the post form

$$\langle \beta | \hat{T}_{\alpha\beta}^{(po)} | \alpha \rangle = \langle \beta | \hat{V}^{(\beta)} | \alpha, + \rangle, \qquad (7.20)$$

one obtains in a similar way

$$\hat{T}_{\alpha\beta}^{(po)} = \hat{V}^{(\beta)} + \hat{V}^{(\beta)} \hat{G}^{(+)}(E) \hat{V}^{(\alpha)}. \qquad (7.21)$$

Comparison of (7.19) and (7.21) shows that two different expressions exist for the T-matrix operator, which characterise the transition from channel α into the channel

β. However, the chain of transformations

$$\langle \beta K'|(\hat{\mathsf{T}}_{\alpha\beta}^{(po)} - \hat{\mathsf{T}}_{\alpha\beta}^{(pr)})|\alpha K\rangle = \langle \beta K'|(\hat{V}^{(\beta)} - \hat{V}^{(\alpha)})|\alpha K\rangle$$

$$= \langle \beta K'|(\hat{H}^{(\alpha)} - \hat{H}^{(\beta)})|\alpha K\rangle = (E(\alpha, K) - E(\beta, K'))\langle \beta K'|\alpha K\rangle$$

$$= 0 \quad \text{for} \quad E(\alpha, K) = E(\beta, K')$$

demonstrates, that they lead to the same on-shell matrix element.

Explicit integral equations for the T-matrix operators are obtained in the following fashion:

- Prior form: Multiplication of (7.19) from the *left* with $\hat{G}_\beta^{(+)}(E)$ gives with the Dyson equation (7.15) for the Green's function the expression

$$\hat{G}_\beta^{(+)}(E)\hat{\mathsf{T}}_{\alpha\beta}^{(pr)} = \left(\hat{G}_\beta^{(+)}(E) + \hat{G}_\beta^{(+)}(E)\hat{V}^{(\beta)}\hat{G}^{(+)}(E)\right)\hat{V}^{(\alpha)}$$

$$= \hat{G}^{(+)}(E)\hat{V}^{(\alpha)}.$$

Replacement of the product $\hat{G}\hat{V}$ in (7.19) with this result leads to the integral equation

$$\hat{\mathsf{T}}_{\alpha\beta}^{(pr)} = \hat{V}^{(\alpha)} + \hat{V}^{(\beta)}\hat{G}_\beta^{(+)}(E)\hat{\mathsf{T}}_{\alpha\beta}^{(pr)}. \tag{7.22}$$

- Post-form: Multiplication of (7.21) from the *right* with $\hat{G}_\alpha^{(+)}(E)$ yields with (7.16)

$$\hat{\mathsf{T}}_{\alpha\beta}^{(po)}\hat{G}_\alpha^{(+)}(E) = \hat{V}^{(\beta)}\left(\hat{G}_\alpha^{(+)}(E) + \hat{G}^{(+)}(E)\hat{V}^{(\alpha)}\hat{G}_\alpha^{(+)}(E)\right)$$

$$= \hat{V}^{(\beta)}\hat{G}^{(+)}(E).$$

Substitution of this result into (7.21) gives

$$\hat{\mathsf{T}}_{\alpha\beta}^{(po)} = \hat{V}^{(\beta)} + \hat{\mathsf{T}}_{\alpha\beta}^{(po)}\hat{G}_\alpha^{(+)}(E)\hat{V}^{(\alpha)}. \tag{7.23}$$

In addition, one can obtain relations which relate the T-matrix operators $\hat{\mathsf{T}}_{\alpha\beta}^{(pr)}$ and $\hat{\mathsf{T}}_{\alpha\beta}^{(po)}$ with the operator

$$\hat{\mathsf{T}}_{\alpha\alpha} = \hat{V}^{(\alpha)} + \hat{V}^{(\alpha)}\hat{G}^{(+)}(E)\hat{V}^{(\alpha)}$$

and with $\hat{\mathsf{T}}_{\beta\beta}$. For this purpose, one considers

$$\hat{G}_\alpha^{(+)}(E)\hat{\mathsf{T}}_{\alpha\alpha} = \hat{G}^{(+)}(E)\hat{V}^{(\alpha)}$$

as well as

$$\hat{T}_{\beta\beta}\hat{G}_{\beta}^{(+)}(E) = \hat{V}^{(\beta)}\hat{G}^{(+)}(E)$$

and obtains

$$\hat{T}_{\alpha\beta}^{(pr)} = \hat{V}^{(\alpha)} + \hat{V}^{(\beta)}\hat{G}_{\alpha}^{(+)}(E)\hat{T}_{\alpha\alpha} \tag{7.24}$$

as well as

$$\hat{T}_{\alpha\beta}^{(po)} = \hat{V}^{(\beta)} + \hat{T}_{\beta\beta}\hat{G}_{\beta}^{(+)}(E)\hat{V}^{(\alpha)}. \tag{7.25}$$

Similar to the case of pure elastic scattering, the integral equations for the T-matrix operators are suitable for the formulation of approximations. The simplest is the Born approximation

$$\langle \beta K' | \hat{T}_{\alpha\beta} | \alpha K \rangle_{\text{Born}} = \langle \beta K' | \hat{V}^{(\alpha)} | \alpha K \rangle = \langle \beta K' | \hat{V}^{(\beta)} | \alpha K \rangle.$$

As an illustration of possible results in this approximation, one can look at

- elastic scattering $a + (bc) \rightarrow a + (bc)$ and
- the excitation of the target $a + (bc) \rightarrow a + (bc)^*$

for the case that c is a heavy particle, so that the position of c can be chosen as the origin of the coordinate system. In consequence, instead of $r_{ac} = r_a - r_c$ and $r_{bc} = r_b - r_c$ only the coordinates r_a and r_b are required. The interaction in the two channels is

$$\hat{V}^{(1)} + \hat{V}^{(2)} \rightarrow \hat{V}_{(ac)}(r_{ac}) + \hat{V}_{(ab)}(r_{ab}) \rightarrow v_{(ac)}(r_a) + v_{(ab)}(r_{ab}).$$

The asymptotic wave function for elastic scattering is (assuming a short-range interaction)

$$\langle r_a, r_b | 1, k \rangle = \frac{1}{(2\pi)^{3/2}} e^{ik \cdot r_a} \phi_0(r_b).$$

The first factor represents the (free) motion of a relative to c, the second factor the binding of b to c in the ground state. The T-matrix element in the Born approximation is

$$\langle 1, k' | \hat{T}_{11} | 1, k \rangle_{\text{Born}} = \frac{1}{(2\pi)^3} \int d^3 r_a \int d^3 r_b e^{-ik' \cdot r_a} \phi_0^*(r_b)$$
$$\times \left[v_{(ac)}(r_a) + v_{(ab)}(r_{ab}) \right] e^{ik \cdot r_a} \phi_0(r_b).$$

The first term describes the elastic scattering of a by the heavy particle

$$\langle 1, \mathbf{k}' | \hat{T}_{11} | 1, \mathbf{k} \rangle_{B1} = \frac{1}{(2\pi)^3} \int d^3 r_a v_{(ac)}(\mathbf{r}_a) e^{i\mathbf{q} \cdot \mathbf{r}_a} = v_{(ac)}(\mathbf{q})$$

via the Fourier transform of the scattering potential as a function of the momentum transfer $\hbar \mathbf{q} = \hbar(\mathbf{k} - \mathbf{k}')$. The second term can be evaluated as follows: One writes the integral in the form

$$\langle 1, \mathbf{k}' | \hat{T}_{11} | 1, \mathbf{k} \rangle_{B2} = \frac{1}{(2\pi)^3} \int d^3 r_{ab} v_{(ab)}(\mathbf{r}_{ab}) e^{i\mathbf{q} \cdot \mathbf{r}_{ab}} \int d^3 r_b e^{i\mathbf{q} \cdot \mathbf{r}_b} |\phi_0(\mathbf{r}_b)|^2$$

$$= v_{(ab)}(\mathbf{q}) F_{00}(\mathbf{q})$$

after transforming the integrand with

$$\exp(i\mathbf{q} \cdot \mathbf{r}_b)\exp(-i\mathbf{q} \cdot \mathbf{r}_b) = 1.$$

The Fourier transform of the potential $v_{(ab)}$ expresses the elastic scattering of a by the *free* particle b. The fact that b is not free is expressed by the second factor, the Fourier transform of the probability distribution of the particle b in the ground state. The quantity F_{00} is called the *elastic form factor*. The measurement of the form factor as a function of momentum transfer provides, within the Born approximation, information about the spatial distribution of b in the system (bc), i.e. the target structure.

If the target is excited, the asymptotic wave function of the final state is

$$\langle \mathbf{r}_a, \mathbf{r}_b | 2, \mathbf{k} \rangle = \frac{1}{(2\pi)^{3/2}} e^{i\mathbf{k} \cdot \mathbf{r}_a} \phi_i(\mathbf{r}_b) \qquad i = 1, 2, \ldots \quad .$$

As a result of the orthogonality of the wave functions of the ground state and excited states, only the term with the interaction $v_{(ab)}$ contributes, so that one obtains

$$\langle 2, \mathbf{k}' | \hat{T}_{12} | 1, \mathbf{k} \rangle_B = \frac{1}{(2\pi)^3} \int d^3 r_{ab} v_{(ab)}(\mathbf{r}_{ab}) e^{i\mathbf{q} \cdot \mathbf{r}_{ab}}$$

$$\times \int d^3 r_b e^{i\mathbf{q} \cdot \mathbf{r}_b} \phi_i^*(\mathbf{r}_b) \phi_0(\mathbf{r}_b) = v_{(ab)}(\mathbf{q}) F_{0i}(\mathbf{q}).$$

Instead of an elastic form factor, one finds an inelastic form factor. It should also be noted that the wave numbers of the two states must be different.

The question of the prefactors that appear in final expressions for differential cross sections is addressed only briefly. They feature as factors of the square modulus of the T-matrix. In general, one can assume that there are only two collision partners in the initial channel. The mass factors are given by the reduced mass μ, the momentum factors (or wave numbers) are determined by the relative momenta in the laboratory system of the respective channels, e.g. for the entrance channel $a + (bc)$

$$\mu_\alpha = \frac{m_a m_{(bc)}}{m_a + m_{(bc)}},$$

$$\kappa_\alpha = \frac{m_{(bc)} \mathbf{k}_a - m_a \mathbf{k}_{(bc)}}{m_a + m_{(bc)}}.$$

For an exit channel β with two collision partners, the differential cross section is

$$\left(\frac{d\sigma}{d\Omega}\right)_{\alpha \to \beta} = \frac{(2\pi)^4}{\hbar^4} \mu_\alpha \mu_\beta \frac{\kappa_\beta}{\kappa_\alpha} |\langle \beta | \hat{T}_{\alpha\beta} | \alpha \rangle|^2.$$

In the case of elastic scattering one has $\mu_\alpha = \mu_\beta = \mu$ and $k_\alpha = k_\beta = k$, so that one recovers the form used in the earlier chapters.

For final channels with three particles one can use two sets of relative coordinates in the form of vectors and corresponding relative momenta. A possible choice of the coordinates in this case are hyperspherical coordinates[3] or Jacobi coordinates.[4]

7.5   The Faddeev Equations

It was noted in Sect. 7.3, that the integral equations for the three-body collision problem, which were derived in that chapter, can not be used directly. A practicable formulation of the three-body collision problem was given by Faddeev.[5] The collision system, which will be considered in this context, involves again three distinguishable particles $(a, b, c) \equiv (1, 2, 3)$ with the possible channels:

(0) (a, b, c) move freely.
(1) a is free, (bc) are bound.
(2) b is free, (ac) are bound.
(3) c is free, (ab) are bound.
(4) The possible channel in which (abc) are bound is not considered with the reference to the energy situation.

The interactions in the four channels 0 to 3 can either

- be characterised, as in Sect. 7.1, by a channel number

---

[3] See, e.g., L. E. Espinola Lopez and J. J. Soares Neto, Int. J. Theor. Phys. **39**, p. 1129 (2000).

[4] See, for example, A. G. Sitenko, Lectures in Scattering Theory. Pergamon Press, Oxford (1971), p. 192.

[5] L. D. Faddeev, JETP. **12**, p. 1014 (1961), C. Lovelace, Phys. Rev. **135**, p. B1225 (1964) see also W. Glöckle: The Quantum Mechanical Few-Body Problem. Springer Verlag, Heidelberg (1983).

- or, as in Sect. 7.3, by the explicit interactions present between the colliding particles (using the notation 1, 2, 3)

$$
\begin{aligned}
&\text{channel} && \text{explicit} \\
\hat{V}^{(0)} &= \quad \hat{v}_{(12)} + \hat{v}_{(23)} + \hat{v}_{(13)}, \\
\hat{V}^{(1)} &= \quad \hat{v}_{(12)} + \hat{v}_{(13)}, \\
\hat{V}^{(2)} &= \quad \hat{v}_{(12)} + \hat{v}_{(23)}, \\
\hat{V}^{(3)} &= \quad \hat{v}_{(23)} + \hat{v}_{(13)}.
\end{aligned}
$$

The necessity for an additional investigation of the Lippmann-Schwinger equations for the T-matrix elements, which have been treated in the previous section, arises from the fact that these equations lead to singular T-matrix elements and hence to singular cross sections. This can be demonstrated in a direct way by using the prior form of the T-matrix equation (7.22) for the case of a transition[6] from channel 0 to channel 0. The appropriate T-matrix operator[7] is given by the equation

$$
\hat{T}_{00} = \hat{V}^{(0)} + \hat{V}^{(0)} \hat{G}_0^{(+)} \hat{T}_{00} \tag{7.26}
$$

with

$$
\hat{G}_0^{(+)} = (E + i\epsilon - \hat{T}_1 - \hat{T}_2 - \hat{T}_3)^{-1}.
$$

Iteration of Eq. (7.26) in diagrammatic form yields for the matrix element

$$
\langle 0, \, \mathbf{k'}_1, \mathbf{k'}_2, \mathbf{k'}_3 | \hat{T}_{00} | 0, \, \mathbf{k}_1, \mathbf{k}_2, \mathbf{k}_3 \rangle
$$

in the momentum representation the result, which is indicated in Fig. 7.1. The perturbation series contains subdiagrams, in which one of the particles does not interact with the other particles. The resummation of these disconnected diagrams, as illustrated in Fig. 7.2, indicates, that the sum

$$
\langle \mathbf{k'}_1, \mathbf{k'}_2, \mathbf{k'}_3 | \hat{T}_{00} | \mathbf{k}_1, \mathbf{k}_2, \mathbf{k}_3 \rangle_{\text{partial}} = \delta(\mathbf{k'}_1 - \mathbf{k}_1) \langle \mathbf{k'}_2, \mathbf{k'}_3 | \hat{t}_{12} | \mathbf{k}_2, \mathbf{k}_3 \rangle
$$

leads to δ-functions and thus ultimately to ill-conditioned integral kernels in the Lippmann-Schwinger equations, as for instance in Eq. (7.26). The elimination of these divergencies can be achieved by resummation of certain classes of contribu-

---

[6] In an experiment, this may not be easy to implement, as three beams have to be crossed in the entrance channel.

[7] The arguments of the next steps are based on the prior form.

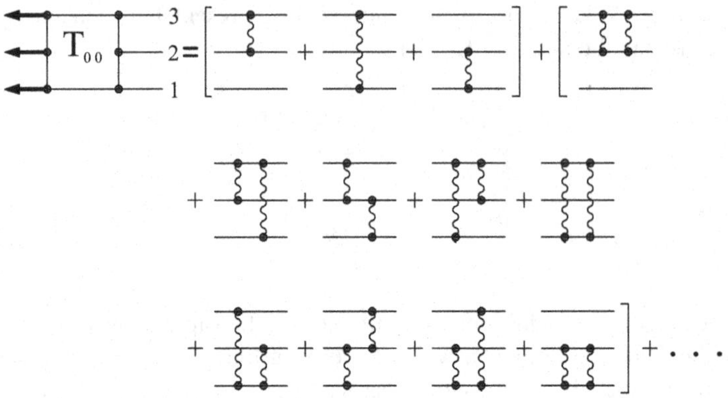

Fig. 7.1 Perturbation expansion of the matrix element of the operator $\hat{T}_{00}$

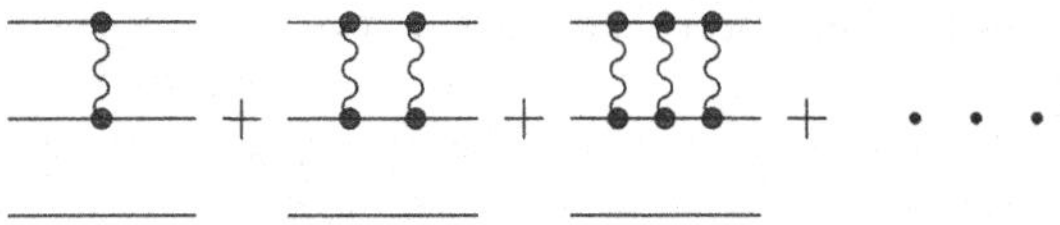

Fig. 7.2 Partial resummation of a divergent contribution

tions of the direct perturbation expansion. In order to arrive at this rearrangement of the perturbation expansion, Faddeev proposed the following procedure:

The channel Lippmann-Schwinger equation (7.17) for the entrance channel

$$|\alpha, +\rangle = |\alpha\rangle + \hat{G}_\alpha^{(+)}(E(\alpha)) \, \hat{V}^{(\alpha)} \, |\alpha, +\rangle$$

was derived from Eq. (7.10)

$$|\alpha, +\rangle = |\alpha\rangle + \hat{G}^{(+)}(E(\alpha)) \, \hat{V}^{(\alpha)} \, |\alpha\rangle$$

by applying the general relation (7.15)

$$\hat{G}^{(+)}(E) = \hat{G}_\gamma^{(+)}(E) + \hat{G}_\gamma^{(+)}(E)\hat{V}^{(\gamma)}\hat{G}^{(+)}(E)$$

for $\gamma = \alpha$. A rearrangement of the perturbation series can be obtained by replacing the exact Green's function in (7.10) by a channel Green's function in a channel $\beta \neq \alpha$ using (7.15). One first finds (with $E(\alpha) = E(\beta) = E$, as the total energy is the same in all channels)

$$\hat{G}^{(+)}(E)\,\hat{V}^{(\alpha)}\,|\alpha\rangle = \left\{\hat{G}_\beta^{(+)}(E)\,\hat{V}^{(\alpha)} + \hat{G}_\beta^{(+)}(E)\,\hat{V}^{(\beta)}\hat{G}^{(+)}(E)\,\hat{V}^{(\alpha)}\right\}|\alpha\rangle$$
$$=\hat{G}_\beta^{(+)}(E)(\hat{V}^{(\alpha)} - \hat{V}^{(\beta)})|\alpha\rangle + \hat{G}_\beta^{(+)}(E)\,\hat{V}^{(\beta)}|\alpha, +\rangle.$$

If one inserts this expression into the α-channel Lippmann-Schwinger equation (7.10), one obtains with an extension

$$|\alpha, +\rangle =\hat{G}_\beta^{(+)}(E)(E - \hat{H}^{(\beta)} + i\epsilon + \hat{V}^{(\alpha)} - \hat{V}^{(\beta)})|\alpha\rangle + \hat{G}_\beta^{(+)}(E)\,\hat{V}^{(\beta)}|\alpha, +\rangle.$$

The factor of $\hat{G}_\beta^{(+)}$ in the first term reduces to

$$(E - \hat{H}^{(\beta)} + i\epsilon + \hat{V}^{(\alpha)} - \hat{V}^{(\beta)})|\alpha\rangle = (E + i\epsilon - \hat{H}^{(\alpha)})|\alpha\rangle = i\epsilon|\alpha\rangle$$

as the definition of the channel potential in Eq. (7.7) is

$$\hat{V}^{(\gamma)} = \hat{H} - \hat{H}^{(\gamma)}.$$

The proof that the limit

$$\lim_{\epsilon\to 0} i\epsilon\hat{G}_\beta^{(+)}(E)|\alpha\rangle$$

has the value zero (known as *Lippmann's identity*), if β is not equal to α, is given in Sect. 7.7.3. Thus, one is left with the statement, that the equations

$$|\alpha, +\rangle = \lim_{\epsilon\to 0} \hat{G}_\beta^{(+)}(E(\alpha))\,\hat{V}^{(\beta)}|\alpha, +\rangle \quad \text{for} \quad \beta \neq \alpha \tag{7.27}$$

can be derived for the entrance channel α in addition to the Lippmann-Schwinger equation for this channel. The consequence is:

- Without the additional homogeneous Eq. (7.27) one does not obtain a unique solution of the entrance channel Lippmann-Schwinger equation (7.10) in the case of multichannel problems.
- In order to obtain a unique solution, one has to ascertain, that in addition to the solution of the inhomogeneous Lippmann-Schwinger equation (7.10) the homogeneous equations (7.27)

$$|\alpha, +\rangle = \hat{G}_\beta^{(+)}(E)\,\hat{V}^{(\beta)}|\alpha, +\rangle \quad \text{where} \quad \beta \neq \alpha$$

is implemented. As the states $|\beta, +\rangle$ satisfy the inhomogeneous equations

$$|\beta, +\rangle = |\beta\rangle + \hat{G}_\beta^{(+)}(E)\,\hat{V}^{(\beta)}|\beta, +\rangle$$

the implementation of this requirement excludes them as components in the representation of $|\alpha, +\rangle$.

The problem with four channels 0, 1, 2, 3, stated above, can be used to illustrate the explicit steps. The entrance channel, in which the particle a collides with the bound system (bc) is channel 1. The T-matrix elements of interest are therefore

$$\langle \beta | \hat{T}_{1\beta} | 1 \rangle \quad \beta = 0, \ldots, 3.$$

It will be shown below, that only the states representing the elastic and the transfer channels need to be included in the calculation, as the T-matrix element for the breakup channel $\langle 0 | \hat{T}_{10} | 1 \rangle$ can be obtained from the T-matrix elements of the channels with $\beta = 1, 2, 3$. It is also convenient to use the cyclic notation

$$\hat{v}_1 = \hat{v}_{(23)}, \quad \hat{v}_2 = \hat{v}_{(13)}, \quad \hat{v}_3 = \hat{v}_{(12)}.$$

The Lippmann-Schwinger equation (7.17) and the additional conditions (7.27) are then [8]

$$
\begin{aligned}
|1, +\rangle &= |1\rangle + \hat{G}_1^{(+)}(E)(\hat{v}_2 + \hat{v}_3)|1, +\rangle, \\
|1, +\rangle &= \qquad \hat{G}_2^{(+)}(E)(\hat{v}_3 + \hat{v}_1)|1, +\rangle, \qquad (7.28) \\
|1, +\rangle &= \qquad \hat{G}_3^{(+)}(E)(\hat{v}_1 + \hat{v}_2)|1, +\rangle.
\end{aligned}
$$

On the right-hand side of (7.28), one recognises the T-matrix operators for the elastic channel ($\hat{T}_{11}$) and the transfer channels ($\hat{T}_{12}$ and $\hat{T}_{13}$). According to (7.20) one has

$$
\begin{aligned}
\hat{T}_{11}|1\rangle &= (\hat{v}_2 + \hat{v}_3)|1, +\rangle, \\
\hat{T}_{12}|1\rangle &= (\hat{v}_3 + \hat{v}_1)|1, +\rangle, \\
\hat{T}_{13}|1\rangle &= (\hat{v}_1 + \hat{v}_2)|1, +\rangle,
\end{aligned}
$$

so that the system of equations (7.28) can be sorted in the form

$$
\begin{aligned}
\hat{T}_{11}|1\rangle &= \left[\hat{v}_2 \hat{G}_2^{(+)}(E)\hat{T}_{12} + \hat{v}_3 \hat{G}_3^{(+)}(E)\hat{T}_{13} \right]|1\rangle, \\
\hat{T}_{12}|1\rangle &= \left[\hat{v}_3 \hat{G}_3^{(+)}(E)\hat{T}_{13} + \hat{v}_1 + \hat{v}_1 \hat{G}_1^{(+)}(E)\hat{T}_{11} \right]|1\rangle, \qquad (7.29) \\
\hat{T}_{13}|1\rangle &= \left[\hat{v}_1 + \hat{v}_1 \hat{G}_1^{(+)}(E)\hat{T}_{11} + \hat{v}_2 \hat{G}_2^{(+)}(E)\hat{T}_{12} \right]|1\rangle.
\end{aligned}
$$

---

[8] The argumentation uses the post form.

The asymptotic state $|1\rangle$ satisfies the Schrödinger equation

$$(\hat{H}_0 + \hat{v}_1)|1\rangle = E|1\rangle,$$

so that one can write the driving term in the equations (7.29) in the form

$$\hat{v}_1|1\rangle = (E - \hat{H}_0|1\rangle = (\hat{G}_0^{(+)}(E))^{-1}|1\rangle.$$

The system of equations for the T-matrix operators obtained after this step ($\beta = 1, ...,$)

$$\hat{T}_{1\beta} = (1 - \delta_{1\beta})(\hat{G}_0^{(+)}(E))^{-1} + \sum_{\gamma=1}^{3}(1 - \delta_{\beta\gamma})\hat{v}_\gamma \hat{G}_\gamma^{(+)}(E)\hat{T}_{1\gamma} \qquad (7.30)$$

is known as the Alt-Grassberger-Sandhas system [9].

A variant of (7.29), respectively of (7.30), is obtained by introducing the T-matrix operators of the *two-particle subsystems* $\hat{t}_\gamma$ in the three-particle space. They satisfy the Lippmann-Schwinger equation

$$\hat{t}_\gamma = \hat{v}_\gamma + \hat{v}_\gamma \hat{G}_0^{(+)}(E)\hat{t}_\gamma, \quad \gamma = 1, 2, 3,$$

with the free Green's function of the *three-particle problem*

$$\hat{G}_0^{(+)}(E) = (E + i\epsilon - \hat{T}_1 - \hat{T}_2 - \hat{T}_3)^{-1}.$$

The equation for these operators results from the definition

$$\hat{v}_\gamma \hat{G}_\gamma^{(+)}(E) = \hat{t}_\gamma \hat{G}_0^{(+)}(E) \qquad (7.31)$$

and the relation

$$\hat{G}_\gamma^{(+)}(E) = \hat{G}_0^{(+)}(E) + \hat{G}_0^{(+)}(E)\hat{v}_\gamma \hat{G}_\gamma^{(+)}(E).$$

If one introduces these operators into the system of equations (7.30), one arrives at the *Faddeev equations*

$$\hat{T}_{1\beta} = (1 - \delta_{1\beta})v_1 + \sum_{\gamma=1}^{3}(1 - \delta_{\beta\gamma})\hat{t}_\gamma \hat{G}_0^{(+)}(E)\hat{T}_{1\gamma} \quad \beta = 1, ..., 3 \quad . \qquad (7.32)$$

---

[9] E. O. Alt, P. Grassberger, W. Sandhas, Nucl. Phys. **B2**, p. 167 (1967).

The equivalent matrix equation has the structure

$$
\begin{pmatrix} \hat{T}_{11} \\ \hat{T}_{12} \\ \hat{T}_{13} \end{pmatrix} = \begin{pmatrix} \hat{0} \\ \hat{v}_1 \\ \hat{v}_1 \end{pmatrix} + \begin{pmatrix} \hat{0} & \hat{t}_2 & \hat{t}_3 \\ \hat{t}_1 & \hat{0} & \hat{t}_3 \\ \hat{t}_1 & \hat{t}_2 & \hat{0} \end{pmatrix} \hat{G}_0^{(+)}(E) \begin{pmatrix} \hat{T}_{11} \\ \hat{T}_{12} \\ \hat{T}_{13} \end{pmatrix}.
\tag{7.33}
$$

The first iteration of this system of equations

$$
\begin{pmatrix} (\hat{T}_{11})_{[1]} \\ (\hat{T}_{12})_{[1]} \\ (\hat{T}_{13})_{[1]} \end{pmatrix} = \begin{pmatrix} \hat{t}_2 \hat{G}_0^{(+)} \hat{v}_1 + \hat{t}_3 \hat{G}_0^{(+)} \hat{v}_1 \\ \hat{t}_3 \hat{G}_0^{(+)} \hat{v}_1 \\ \hat{t}_2 \hat{G}_0^{(+)} \hat{v}_1 \end{pmatrix}
$$

shows, that as a consequence of the reorganisation of the terms no disconnected diagrams occur (except for the trivial terms in zeroth order).

The T-matrix element for the breakup channel $\langle 0|\hat{T}_{10}|1\rangle$ can be expressed with (7.25) by the T-matrix elements of the other channels. The starting point for the demonstration of this statement is

$$
\langle 0|\hat{T}_{10}|1\rangle = \sum_{\gamma=1}^{3} \langle 0|\hat{v}_\gamma|1, +\rangle.
$$

From Eq. (7.28) one can extract the relation

$$
\langle 0|\hat{T}_{10}|1\rangle = \sum_{\gamma=1}^{3} \langle 0|v_\gamma \hat{G}_\gamma^{(+)}(E)\hat{T}_{1\gamma}|1\rangle
$$

$$
= \sum_{\gamma=1}^{3} \langle 0|t_\gamma \hat{G}_0^{(+)}(E)\hat{T}_{1\gamma}|1\rangle.
$$

In this step the on-shell equation

$$
\langle 0|v_1|1\rangle = \langle 0|\hat{H}^{(1)} - \hat{H}^{(0)}|1\rangle = (E - E)\langle 0||1\rangle = 0
$$

has been used.

The results obtained can be generalised without additional calculations. If the initial state is indexed by α as before, the equations for the T-matrix operators are instead of (7.30)

$$
\hat{T}_{\alpha\beta} = (1 - \delta_{\alpha\beta})(\hat{G}_0^{(+)}(E))^{-1} + \sum_{\gamma}(1 - \delta_{\alpha\gamma})\hat{v}_\gamma \hat{G}_\gamma^{(+)}(E)\hat{T}_{\alpha\gamma}.
\tag{7.34}
$$

In Faddeev's original work the states characterising the channels were analysed in more detail rather than the T-matrix operators. For the present example the decomposition

$$|\alpha, +\rangle = \sum_{\mu=1}^{3} \hat{G}_0^{(+)}(E)\hat{v}_\mu|\alpha, +\rangle \qquad (7.35)$$

can be used to rewrite the Schrödinger equation

$$\left(\hat{H}_0 + \sum_{\mu=1}^{3} \hat{v}_\mu\right)|\Psi\rangle = E|\Psi\rangle$$

as an integral equation

$$|\Psi\rangle = \left(\frac{1}{E - \hat{H}_0 + i\epsilon}\right) \sum_{\mu=1}^{3} \hat{v}_\mu|\Psi\rangle.$$

Iteration yields

$$|\Psi\rangle = \hat{G}_0^{(+)}(E) \sum_{\mu_1=1}^{3} \hat{v}_{\mu_1} \hat{G}_0^{(+)}(E) \sum_{\mu_2=1}^{3} \hat{v}_{\mu_2} \hat{G}_0^{(+)}(E) \sum_{\mu_3=1}^{3} \hat{v}_{\mu_3} \hat{G}_0^{(+)}(E)\ldots|\Psi\rangle.$$

Interactions between all pairs appear in any order. In Faddeev's approach, the interactions between each of the pairs are *first* summed up to order infinity, which amounts to a rearrangement of the direct expansion.

In order to use this ansatz further, one acts with the operators

$$\hat{G}_0^{(+)}(E)\hat{v}_\alpha, \ \hat{G}_0^{(+)}(E)\hat{v}_\beta, \ \hat{G}_0^{(+)}(E)\hat{v}_\gamma$$

on the three equations (generalisation of (7.28))

$$\begin{aligned}
|\alpha, +\rangle &= |\alpha\rangle + \hat{G}_\alpha^{(+)}(E)\hat{V}^{(\alpha)}|\alpha, +\rangle, \\
|\alpha, +\rangle &= \hat{G}_\beta^{(+)}(E)\hat{V}^{(\beta)}|\alpha, +\rangle, \qquad (7.36) \\
|\alpha, +\rangle &= \hat{G}_\gamma^{(+)}(E)\hat{V}^{(\gamma)}|\alpha, +\rangle
\end{aligned}$$

and uses the notation

$$|\alpha, \mu\rangle = \hat{G}_0^{(+)}(E)\hat{v}_\mu|\alpha, +\rangle$$

as well as the rearranged asymptotic Schrödinger equation

$$\hat{G}_0^{(+)}(E)\hat{v}_\alpha|\alpha\rangle = |\alpha\rangle$$

and the relation

$$\hat{G}_0^{(+)}(E)\hat{v}_\alpha\hat{G}_\alpha^{(+)}(E) = \hat{G}_\alpha^{(+)}(E)\hat{v}_\alpha\hat{G}_0^{(+)}(E)$$

in order to find

$$
\begin{aligned}
|\alpha, \alpha\rangle &= |\alpha\rangle &+ \hat{G}_\alpha^{(+)}(E)\hat{v}_\alpha\{|\alpha, \beta\rangle + |\alpha, \gamma\rangle\}, \\
|\alpha, \beta\rangle &= &+ \hat{G}_\beta^{(+)}(E)\hat{v}_\beta\{|\alpha, \gamma\rangle + |\alpha, \alpha\rangle\}, \\
|\alpha, \gamma\rangle &= &+ \hat{G}_\gamma^{(+)}(E)\hat{v}_\gamma\{|\alpha, \alpha\rangle + |\alpha, \beta\rangle\}.
\end{aligned}
\tag{7.37}
$$

With the T-matrix operators for two particles one obtains according to (7.31) a standard form of the Faddeev equations for the states $|\alpha, \mu\rangle$

$$
\begin{pmatrix} |\alpha, \alpha\rangle \\ |\alpha, \beta\rangle \\ |\alpha, \gamma\rangle \end{pmatrix} = \begin{pmatrix} |\alpha\rangle \\ \hat{0} \\ \hat{0} \end{pmatrix} + \hat{G}_0^{(+)}(E) \begin{pmatrix} \hat{0} & \hat{t}_1 & \hat{t}_1 \\ \hat{t}_2 & \hat{0} & \hat{t}_2 \\ \hat{t}_3 & \hat{t}_3 & \hat{0} \end{pmatrix} \begin{pmatrix} |\alpha, \alpha\rangle \\ |\alpha, \beta\rangle \\ |\alpha, \gamma\rangle \end{pmatrix}.
\tag{7.38}
$$

The matrix formed from the T-matrices is the mirror image of the matrix in (7.33). This suggests a connection between the two strands of the argumentation. It is possible to derive (7.33) with (7.38) in a different way.

The structure of these matrices with the value zero for the diagonal elements is the reason why the multiple scattering expansions lead to reasonable results. The first iteration of the kernel of the integral equations yields a Schmidt-Hilbert type kernel, which admits only connected diagrams.

For the actual application of the Faddeev formulation, appropriate coordinates in ordinary space or in momentum space have to be selected. Exchange symmetry has to be included if required. It may also be remarked that the method developed by Faddeev (and others) has led to many variants (which are not addressed here) and that it can be applied in principle to systems with more than three particles.

7.6   Transfer Reactions

7.6.1   A Short Note on Nuclear Reactions

One of the topics that involves *multichannel problems* is the theory of nuclear reactions. It will be introduced in this section by a brief discussion of the example of (d,p)-transfer reactions. A point of particular interest is the implementation of the Born approximation with modified waves, the DWBA (see Sect. 2.2.1). The deuteron projectile in the reaction

$$^{208}_{82}\text{Pb} + ^2_1\text{H} \longrightarrow ^{209}_{82}\text{Pb} + ^1_1\text{H},$$

which is written in short form as

$$^{208}\text{Pb (d, p) } ^{209}\text{Pb}, \qquad (7.39)$$

is relatively weakly bound, so that it is readily decomposed into its components. The neutron is captured by the lead nucleus ^{208}Pb, forming the lead isotope 209. The proton is repelled, as it has the same sign of charge as the lead nucleus, and escapes. Two entirely different scenarios can be imagined for this reaction:

- The reaction is called *direct* or a direct process, if the capture of the neutron takes place without further intermediate steps, for example at the nuclear surface. A measure for the duration of the reaction is the flyby time of the projectile

$$t_{\text{dir}} = \frac{2R_{\text{nucleus}}}{v_{\text{proj}}} \approx 10^{-22}s.$$

The number given corresponds to a nuclear radius R_{nucleus} of the order of 10^{-12} cm and a kinetic energy of the deuteron of 100 MeV (corresponding to a velocity $v_{\text{proj}} \approx 10^{10}\text{cm/s}$).

- The reaction is called a *compound process*, if the captured neutron interacts several times with the nucleons in the nucleus so that it transfers most of its energy to the nucleus. The time required for the formation of the compound nucleus is on the average larger than the time required for the direct process.

It is, however, not possible to separate the two different processes on the basis of the time scale. The processes take place simultaneously, along with elastic scattering. Some of the beam particles follow the direct reaction path, and some contribute to the formation of compound nuclei. A formal separation is possible on the basis of the theoretical treatment. The total state vector $|\Psi^{(+)}\rangle$ of the collision system can be decomposed with the help of two self-adjoint projection operators

$$\hat{P} = \hat{P}^\dagger \qquad \text{and} \qquad \hat{Q} = \hat{Q}^\dagger$$

with the properties

$$\hat{P} + \hat{Q} = \hat{1}, \quad \hat{P}^2 = \hat{P}, \quad \hat{Q}^2 = \hat{Q}, \quad \hat{P}\hat{Q} = \hat{Q}\hat{P} = \hat{0}.$$

The sum of the two operators in the first statement encompasses the entire Hilbert space of the problem. The second statement identifies the two operators as projection operators, and the third guarantees a complete separation of the two sectors. The Schrödinger equation

$$(E - \hat{H})|\Psi^{(+)}\rangle = 0$$

can be decomposed with the two projection operators into a set of coupled equations for the segments $\hat{P}|\Psi^{(+)}\rangle = |\Psi_P^{(+)}\rangle$ and $\hat{Q}|\Psi^{(+)}\rangle = |\Psi_Q^{(+)}\rangle$ of the Hilbert space generated by the Hamiltonian $\hat{H}$

$$(E - \hat{H}_{PP})|\Psi_P^{(+)}\rangle = \hat{H}_{PQ}|\Psi_Q^{(+)}\rangle,$$
$$(E - \hat{H}_{QQ})|\Psi_Q^{(+)}\rangle = \hat{H}_{QP}|\Psi_P^{(+)}\rangle.$$

A state like $|\Psi_P^{(+)}\rangle$ spans a subspace, for example the space of states characterising direct processes. The state $|\Psi_Q^{(+)}\rangle$ comprises all other processes, including compound processes. Operators as $\hat{H}_{PP}$ act only in the respective subspace, operators like $\hat{H}_{PQ}$ connect the two spaces.

7.6.2 The Born Approximation

This formal separation fails, however, in practice because of the difficulty to not only specify but also to apply the operators that mediate the transitions for the many-particle situation at hand. In order to obtain a hint about ways out of these difficulties, it is useful to consider the simplest approximation. This is the PWBA, the Born approximation with plane waves. If one uses the centre of mass system for a reaction of the form (7.39) with $(A + 2)$ nucleons

$$d + A \longrightarrow p + (A + n),$$

the initial and final states are

$$|i\rangle = |\boldsymbol{k}_d\rangle \, |\phi_d\rangle \, |\Phi_A\rangle,$$
$$|f\rangle = |\boldsymbol{k}_p\rangle \, |\Phi_{(A+1)}\rangle.$$

The interaction, which is responsible for the transition between these states, is the interaction between the neutron and the proton

$$V(|\boldsymbol{r}_n - \boldsymbol{r}_p|) = V(r).$$

The T-matrix element of this interaction

$$\langle f|\hat{T}|i\rangle^*_{\text{Born}} = \int \mathrm{d}^3 r_1 \ldots \mathrm{d}^3 r_A \, \mathrm{d}^3 r_p \, \mathrm{d}^3 r_n \, \mathrm{e}^{(-\mathrm{i}\boldsymbol{k}_d \cdot \boldsymbol{r}_d)} \mathrm{e}^{(\mathrm{i}\boldsymbol{k}_p \cdot \boldsymbol{r}_p)}$$

$$\phi_d^*(r)\Phi_A^*(1 \ldots A)V(r)\Phi_{(A+1)}(1 \ldots A, n)$$

Fig. 7.3 Choice of the coordinates for the (d, p) reaction

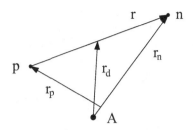

can be calculated in a simple fashion if one assumes that the wave functions of the states of the two nuclei can be represented in terms of the nuclear shell model[10]. The integral over the A coordinates of the nuclear particles then yields the shell model wave function of the neutron

$$\int d^3r_1 \ldots d^3r_A \, \Phi_A^*(1\ldots A)\Phi_{(A+1)}(1\ldots A,\, n) = \phi_{njl}(\boldsymbol{r}_n).$$

The T-matrix in the Born approximation is therefore

$$\langle f|\hat{T}|i\rangle_{\text{Born}}^* = \int d^3r_p \, d^3r_n \, e^{(-i\boldsymbol{k}_d\cdot\boldsymbol{r}_d)} e^{(i\boldsymbol{k}_p\cdot\boldsymbol{r}_p)} \phi_d^*(r) V(r) \phi_{njl}(\boldsymbol{r}_n).$$

In order to discuss this expression, it is necessary to sort the coordinates. Figure 7.3 yields the relations

$$\boldsymbol{r}_p = \frac{A}{(A+1)}\boldsymbol{r}_n - \boldsymbol{r},$$

$$\boldsymbol{r}_d = \boldsymbol{r}_n - \frac{1}{2}\boldsymbol{r}.$$

The position of the captured neutron with respect to the nucleus A is given by the vector $\boldsymbol{r}_n$. The vector $\boldsymbol{r}_p$ denotes the position of the proton with respect to the nucleus in the final channel (A+1). The centre of mass of the deuteron with respect to the nucleus A is the endpoint of the vector $\boldsymbol{r}_d$. The distance of the neutron from the proton in the deuteron is described by the vector $\boldsymbol{r}$.

---

[10] See, for example, A. de Shalit and I. Talmi: Nuclear Shell Theory. Academic Press, New York (1963).

For the sum of the exponents of the plane waves in the entrance channel and in the exit channel, one finds

$$-\mathbf{k}_d \cdot \mathbf{r}_d + \mathbf{k}_p \cdot \mathbf{r}_p = -\left(\mathbf{k}_p - \frac{1}{2}\mathbf{k}_d\right) \cdot \mathbf{r} - \left(\mathbf{k}_d - \frac{A}{(A+1)}\mathbf{k}_p\right) \cdot \mathbf{r}_n$$

$$= -\mathbf{K} \cdot \mathbf{r} - \mathbf{q} \cdot \mathbf{r}_n,$$

so that the T-matrix can be written as

$$\langle f|\hat{T}|i\rangle^*_{\text{Born}} = \left[\int \mathrm{d}^3 r \phi_d^*(r) V(r) \mathrm{e}^{(-i\mathbf{K}\cdot\mathbf{r})}\right] \left[\int \mathrm{d}^3 r_n\, \mathrm{e}^{(-i\mathbf{q}\cdot\mathbf{r}_n)} \phi_{njl}(\mathbf{r}_n)\right]. \qquad (7.40)$$

Equation (7.40) is the result of the early stripping theory formulated by Butler.[11] The first factor in the result is a Fourier transform that pertains only to the deuteron. The second is the part that addresses the structure of the states of the nucleons (in the simplest form) in the final nucleus. The fact that one can obtain information about the nuclear structure via the T-matrix element in the differential cross section is the reason for the interest in the (d, p) and other transfer reactions. If one evaluates this expression further, one finds with the partial wave expansion of the plane waves and the standard form of the shell-model wave function

$$\langle f|\hat{T}|i\rangle_{\text{Born}} \propto Y_{l,m}(\theta) \int_0^\infty \mathrm{d}r\, j_l(qr) R_{n,l}(r).$$

From this expression one can extract the value of the angular momentum of the captured neutron directly or determine it by the integral with the Bessel function and the radial function of the neutron. The pattern revealed is confirmed in outline by experimental results, but the agreement is not optimal. One of the points, where one can start to improve the agreement, is a correction of the trajectories of the incoming and the outgoing particles. In descriptive terms, a plane wave describes particles traveling in a straight line toward or away from the target. Besides the form of the trajectory, one must also consider the fact that a significant fraction of the beam particles initiate compound processes and are not available for direct processes. In order to deal with these objections, *distorted* waves generated by optical potentials are used.

7.6.3 Optical Potentials

Optical potentials were introduced in nuclear physics in order to include the elastic scattering of nucleons by nuclei (and of nuclei by nuclei) in the presence of other

---

[11] S. T. Butler, Phys. Rev. **80**, p. 1095 (1950), Proc. R. Soc. (London) **A208**, p. 559 (1951), Phys. Rev. **88**, p. 685 (1952).

open channels in a phenomenological manner. They depend only on the relative coordinate of the two collision partners (the distance of the two centres of mass), which can be calculated from

$$r_{aA} = \frac{1}{M_a} \sum_{i=1}^{a} m_i r_i - \frac{1}{M_A} \sum_{i=a+1}^{A+a} m_i r_i$$

for the case a + A. The operator for the kinetic energy of the relative motion is (in simple notation)

$$T_{aA} = -\frac{\hbar^2}{2\mu_{aA}} \nabla^2_{aA}$$

with the reduced mass

$$\mu_{aA} = \frac{M_a M_A}{M_a + M_A}.$$

The Schrödinger equation for the relative motion

$$(T_{aA} + U_{aA} - E_{aA})\chi_{aA}(r_{aA}) = 0$$

represents not only the elastic scattering of the two nuclei, but also the reduction of the particle flux, if the optical potential U is a complex quantity. In order to prove this statement, one considers the continuity equation

$$-\frac{\hbar^2}{2\mu_{aA}} \left(\chi^*_{aA}[\nabla^2_{aA}\chi_{aA}] - \chi_{aA}[\nabla^2_{aA}\chi^*_{aA}] \right) = \left(U^*_{aA} - U_{aA} \right) \chi^*_{aA}\chi_{aA}$$

and finds with the definitions of current density and density the relation

$$\hbar \nabla_{aA} \cdot j_{aA} = 2\rho_{aA}\text{Im}(U_{aA}).$$

The current density decreases, if the imaginary part of the optical potential is *negative*.

The optical potential must consist of two parts: one part is determined by the effect of the strong but short-range nuclear forces and one part describes the effect of the weak but long-range Coulomb interaction. For such potentials parametrised expressions are used in most cases. The parameters are determined by analysis and fitting of elastic scattering data for the system of interest. When the parameters have been obtained, the potentials are employed for the calculation of T-matrix elements and the cross sections of the transfer reaction channels.

A situation, that is of of interest, is the reaction of light particles with a nucleus. For modeling the contribution of the nuclear forces to the optical potential a global picture of the nucleus in the form of a density distribution of the nucleons, e.g. by a Woods-Saxon distribution,[12] is often used

$$\rho(r) = \frac{\rho_0}{1 + \exp[(r - R)/a]}.$$

The nuclear radius R varies with the number of nucleons A as

$$R = r_0 A^{1/3}, \quad r_0 \approx (1.2 - 1.4) \times 10^{-12} \, \text{cm} .$$

The parameter a determines the thickness of the nuclear surface. With the assumption that the contribution of the nuclear forces to the optical potential has a similar form, the ansatz for a complex optical potential (with six parameters) generated by the nuclear interaction is

$$U(r) = \frac{V}{1 + \exp[(r - R_v)/a_v]} + \frac{iW}{1 + \exp[(r - R_w)/a_w]}.$$

A simple form of the contribution of the Z charged particles in the nucleus and the Z_p charged particles in a point-like projectile to the optical potential is the Coulomb potential of a uniform charge distribution with the radius R_c

$$V_c(r) = \begin{cases} \frac{Z Z_p e^2}{2 R_c} \left[3 - \left(\frac{r}{R_c} \right)^2 \right] & \text{for} \quad r < R_c, \\ \frac{Z_t Z_p e^2}{r} & \text{for} \quad r > R_c. \end{cases}$$

Besides these simple forms of the optical potential, a large number of variants are used, such as:

- a surface form of the imaginary part of the nuclear contribution by means of the derivative of the Woods-Saxon potential,
- terms with additional spin-orbit interaction in the case of nuclei with an odd number of nucleons,
- non-local potentials instead of the most commonly used local potentials.

Additional information can be found, for example, in the book by P. E. Hodgson: The Nucleon Optical Model. World Scientific, Singapore (1994).

---

[12] R. D. Woods and D. S. Saxon, Phys. Rev. **95**, p. 577 (1954).

7.6.4 DWBA: Born Approximation with Distorted Waves

The equations for the calculation of the T-matrix elements in the DWBA can be derived from the information in Sect. 2.2.1 and 7.4. The first step on the way to a two-potential formula is an extension of the potential function $\hat{V}_\beta$ in a channel β by the optical potential $\hat{U}_\beta$ in this channel

$$\hat{V}^{(\beta)} = \hat{V}_1^{(\beta)} + \hat{V}_2^{(\beta)} = \hat{U}_\beta + \left[\hat{V}^{(\beta)} - \hat{U}_\beta\right]. \tag{7.41}$$

Further argumentation can be based on the post or the prior form. The relevant exact T-matrix elements can be found in Sect. 7.4. The starting point for the discussion using the post-form[13] is Eq. (7.14)

$$\langle \beta K' | \hat{\mathsf{T}}_{\alpha\beta}^{(po)} | \alpha K \rangle = \langle \beta K' | \hat{V}^{(\beta)} | \alpha K + \rangle.$$

Insertion of Eq. (7.41) into this expression gives (compare Eq. (2.36))

$$\langle \beta K' | \hat{\mathsf{T}}_{\alpha\beta}^{(po)} | \alpha K \rangle = \langle K' | \hat{U}_{(\beta)} | K \rangle \delta_{\alpha,\beta} + \langle \beta K' | \hat{V}_2^{(\beta)} | \alpha K + \rangle \tag{7.42}$$

as the optical potential is only a function of the distance between the collision partners. The state $|K\rangle$ is a solution of the collision problem for the relative motion in the optical model. The formal Schrödinger equation for the relative motion is

$$(\hat{T}_\beta + \hat{U}_\beta - E(K))|K\rangle = 0,$$

the explicit equation including the specification of the boundary conditions

$$(T_\beta + U_\beta(r) - E(K))\chi_K^{(\pm)}(r) = 0.$$

Eq. (7.42) is the exact two potential form.

One obtains the DWBA formula for a transfer reaction, e.g. for a (d, p) reaction, if one replaces the exact state by an asymptotic state. As the final state is not the same as the initial state, the first term on the right-hand side of Eq. (7.42) is not present. The second term is evaluated with the solutions of the optical potential problem

$$\langle \beta K' | \hat{\mathsf{T}}_{\alpha\beta}^{(po)} | \alpha K \rangle_{\mathrm{DWBA}} = \langle \beta, \chi_\beta^{(-)} | (\hat{V}^{(\beta)} - \hat{U}_\beta) | \alpha, \chi_\alpha^{(+)} \rangle. \tag{7.43}$$

---

[13] Reminder: A state $|\alpha K\rangle$ is an eigenstate of the asymptotic channel Hamiltonian $\hat{H}^{(\alpha)}|\alpha K\rangle = E(\alpha)|\alpha K\rangle$. A state $|\alpha K+\rangle$ is an exact state of the problem that has developed in time from $t = -\infty$ up to $t = 0$, a state $|\beta K-\rangle$ is an exact state that has developed in the time interval from $t = +\infty$ back to $t = 0$.

A corresponding formula can be obtained with the prior form, starting from Eq. (7.13)

$$\langle \beta K' | \hat{T}_{\alpha\beta}^{(pr)} | \alpha K \rangle = \langle \beta K' - | \hat{V}^{(\alpha)} | \alpha K \rangle.$$

The T-matrix elements calculated with the two formulae give, as shown in Sect. 7.4, identical results only for exact states. To what extent the results differ in the case of approximations has to be checked separately.

The final results look as compact as the formulae of the simple Born approximation. However, the evaluation is much more laborious. Among other things, the solution of the wave equations with optical potentials has to be obtained numerically and implemented numerically for the calculation of the transfer T-matrix elements. The nucleons are fermions. Antisymmetrisation of the nucleons is necessary. The coupling of the orbital angular momentum and the spin of the nucleons has to be included.[14]

7.7   Detailed Calculations for Chap. 7

7.7.1   Energy Conservation of the S-Matrix

The commutation relation

$$\hat{H}\hat{\Omega}_{\pm}^{(\alpha)} = \hat{\Omega}_{\pm}^{(\alpha)}\hat{H}^{(\alpha)}$$

is applied to the expression

$$E(\alpha, K)\langle \beta, K' | \hat{S}_{\alpha\beta} | \alpha, K \rangle$$

and gives

$$E(\alpha, K)\langle \beta, K' | \hat{S}_{\alpha\beta} | \alpha, K \rangle =$$

$$\langle \beta, K' | (\hat{\Omega}_{-}^{(\beta)})^{\dagger} \hat{\Omega}_{+}^{(\alpha)} \hat{H}^{(\alpha)} | \alpha, K \rangle = \langle \beta, K' | (\hat{\Omega}_{-}^{(\beta)})^{\dagger} \hat{H} \hat{\Omega}_{+}^{(\alpha)} | \alpha, K \rangle =$$

$$\langle \beta, K' | \hat{H}^{(\beta)} (\hat{\Omega}_{-}^{(\beta)})^{\dagger} \hat{\Omega}_{+}^{(\alpha)} | \alpha, K \rangle = E(\beta, K')\langle \beta, K' | (\hat{\Omega}_{-}^{(\beta)})^{\dagger} \hat{\Omega}_{+}^{(\alpha)} | \alpha, K \rangle.$$

---

[14] These technical aspects, as well as further developments of the theory, can be gleaned from the books cited in Further Reading. The present remarks on the (d, p)-reaction summarise the development of the theory of direct nuclear reactions in the early years. Readers interested in further developments are referred to the proceedings of the annual meetings on nuclear reactions held at *Varenna*. These reports are published by CERN.

From this follows

$$\left[E(\alpha, K) - E(\beta, K')\right] \langle \beta, K' | \hat{S}_{\alpha\beta} | \alpha, K \rangle = 0.$$

The S-matrix element has the value zero if the energy values of the two states are different.

7.7.2 Steps for an Alternative Derivation of the Faddeev-Lovelace Equations

The starting point is, for example, the post form (7.21) of the channel T-matrix operators (with $i, k = 0, 1, \ldots, 3$)

$$\hat{T}_{ik} = \hat{V}^{(k)} \left(\hat{1} + \hat{G}^{(+)}(E) \hat{V}^{(i)} \right). \tag{7.44}$$

The potential $\hat{V}^{(i)}$ is written in cyclic form

$$\hat{V}^{(i)} = \sum_{l=1}^{3} (1 - \delta_{il}) \hat{v}_l \tag{7.45}$$

and replaced by the T-matrix $\hat{t}_l$

$$\hat{v}_l = \hat{t}_l - \hat{v}_l \hat{G}_0^{(+)}(E) \hat{t}_l.$$

It is then found, that the channel T-matrix operator is given by

$$\hat{T}_{ik} = \hat{V}^{(k)} \left(\hat{1} + \sum_{l=1}^{3} (1 - \delta_{il}) \hat{G}^{(+)}(E) [\hat{t}_l - \hat{v}_l \hat{G}_0^{(+)}(E) \hat{t}_l] \right). \tag{7.46}$$

For the term $\hat{G}^{(+)} \hat{t}_l$ the Dyson equation multiplied by $\hat{t}_l$ from the right is used for the Green's function

$$\hat{G}^{(+)} \hat{t}_l = \hat{G}_0^{(+)} \hat{t}_l + \hat{G}^{(+)} \sum_{m=0}^{3} \hat{v}_m \hat{G}_0^{(+)} \hat{t}_l,$$

so that the term

$$\hat{G}^{(+)}(E) \left[\hat{t}_l - \hat{v}_l \hat{G}_0^{(+)}(E) \hat{t}_l \right]$$

becomes

$$\hat{G}^{(+)}(E)\left[\hat{t}_l - \hat{v}_l\hat{G}_0^{(+)}(E)\hat{t}_l\right] = \hat{G}_0^{(+)}(E)\hat{t}_l + \hat{G}^{(+)}(E)\sum_{m=1}^{3}\hat{v}_m\hat{G}_0^{(+)}(E)\hat{t}_l$$
$$- \hat{G}^{(+)}(E)\hat{v}_l\hat{G}_0^{(+)}(E)\hat{t}_l.$$

The last two terms on the right-hand side can be shown to give

$$\hat{G}^{(+)}(E)\sum_{m=1}^{3}(1 - \delta_{lm})\hat{v}_m\hat{G}_0^{(+)}(E)\hat{t}_l,$$

so that one finds with (7.45)

$$\hat{G}^{(+)}(E)\left[\hat{t}_l - \hat{v}_l\hat{G}_0^{(+)}(E)\hat{t}_l\right] = \hat{G}_0^{(+)}(E)\hat{t}_l + \hat{G}^{(+)}(E)\hat{V}^{(l)}\hat{G}_0^{(+)}(E)\hat{t}_l.$$

If one inserts this expression into (7.46), one finds

$$\hat{T}_{ik} = \hat{V}^{(k)}\left(\hat{1} + \sum_{l=0}^{3}(1 - \delta_{il})\left[\hat{1} + \hat{G}^{(+)}(E)\hat{V}^{(l)}\right]\hat{G}_0^{(+)}(E)\hat{t}_l\right).$$

In the second term on the right-hand side one recognises with (7.44) once more the T-matrix operator $\hat{T}_{lk}$. The final result is

$$\hat{T}_{ik} = \hat{V}^{(k)} + \sum_{l=1}^{3}(1 - \delta_{il})\hat{T}_{lk}\hat{G}_0^{(+)}(E)\hat{t}_l. \tag{7.47}$$

7.7.3   Lippmann's Identity

In order to prove the validity of the relation

$$\lim_{\epsilon \to 0}\frac{i\epsilon}{E_\alpha + i\epsilon - \hat{H}^{(\beta)}}|\alpha,\rangle = 0, \quad \beta \neq \alpha$$

one needs the spectral representation of the resolvent $\hat{G}_\beta^{(+)}$. On the basis of the completeness relation (including states for bound particle pairs plus one free particle and states with three free particles)

$$\sum_{n}d^3q_n\,|nq_n\rangle\langle nq_n| + \int\int d^3q\,d^3p\,|q\,p\rangle\langle q\,p| = 1$$

for the solutions of $\hat{H}^{(\beta)}$ one finds

$$\hat{G}_\beta^{(+)}|\alpha\rangle = \sum_n d^3 q_n \, |n\boldsymbol{q}_n\rangle \left(\frac{1}{E_\alpha + i\epsilon - e_n - \frac{\hbar^2 q_n^2}{2m}} \right) \langle n\boldsymbol{q}_n|\alpha\rangle$$

$$+ \int\int d^3 q \, d^3 p \, |\boldsymbol{q}\,\boldsymbol{p}\rangle \left(\frac{1}{E_\alpha + i\epsilon - \frac{\hbar^2 q^2}{2m} - \frac{\hbar^2 p^2}{2m}} \right) \langle \boldsymbol{q}\,\boldsymbol{p}|\alpha\rangle.$$

The assertion follows as a result of the orthogonality of the states and the fact that the on-shell factors are finite. The limiting value

$$\left[\epsilon \hat{G}_\beta^{(+)} \right]_{\epsilon \to 0}$$

is zero.

Literature in Chap. 7

1. L. E. Espinola Lopez, J. J. Soares Neto, Int. J. Theor. Phys. **39**, p. 1129 (2000)
2. A. G. Sitenko: Lectures in Scattering Theory. Pergamon Press, Oxford (1971)
3. L. D. Faddeev, JETP. **12**, p. 1014 (1961)
4. C. Lovelace, Phys. Rev. **135**, p. B1225 (1964).
5. W. Glöckle: The Quantum Mechanical Few-Body Problem. Springer Verlag, Heidelberg (1983)
6. E. O. Alt, P. Grassberger, W. Sandhas, Nucl. Phys. **B2**, p.167 (1967)
7. A. de Shalit and I. Talmi, Nuclear Shell Theory. Academic Press, New York (1963)
8. S. T. Butler, Phys. Rev. **80**, p. 1095 (1950), Proc. R. Soc. (London) **A208**, p. 559 (1951), and Phys. Rev. **88**, p. 685 (1952)
9. R. D. Woods and D. S. Saxon, Phys. Rev. **95**, p. 577 (1954)
10. P. E. Hodgson, The Nucleon Optical Model. World Scientific, Singapore (1994)

Literature

Further Literature

Given the role, that *scattering theory* plays in physics, the literature on this subject is very extensive. However, for an introduction to the field of non-relativistic collision systems, it seems appropriate to limit the list of literature to the standard texts. The list provided here does not include contributions on special systems, neither from the theoretical nor from the experimental point of view. The literature on mathematically oriented questions is also not documented here.

On the theory of scattering:

1. N. F. Mott and H. S. W. Massey
   The Theory of Atomic Collisions.
   Oxford Clarendon Press, Oxford (1933)
   Last edition: Clarendon Press, Oxford (1965)
2. T. Y Wu and T. Ohmura
   Quantum Theory of Scattering.
   Prentice-Hall, Englewood Cliffs, N. J. (1962).
   Last edition: Dover Publications, Mineola, N. Y. (2011).
3. M. L. Goldberger and K. M. Watson
   Collision Theory.
   Wiley, New York (1964)
   Last edition: Dover, Mineola, N. Y. (2004)
4. R. G. Newton
   Scattering Theory of Waves and Particles
   McGraw-Hill Book Company, New York (1966)
   Last edition: Springer Science and Business Media, Heidelberg (2013).
5. L. S. Rodberg and R. M. Thaler
   Introduction to the Quantum Theory of Scattering.
   Academic Press, New York (1967)

© Springer-Verlag GmbH Germany, part of Springer Nature 2022
R. M. Dreizler et al., *Quantum Collision Theory of Nonrelativistic Particles*,
https://doi.org/10.1007/978-3-662-65591-7

6. A. G. Sitenko
   Lectures in Scattering Theory.
   Pergamon Press, Oxford (1971)
   Translated and Edited by P. J. Shepherd
   Last edition: Springer Verlag, Heidelberg (2012)
7. J. R. Taylor
   Scattering Theory, the Quantum Theory of Nonrelativistic Collisions.
   John Wiley, New York (1972)
   Last edition: Dover Publications, Mineola, N. Y. (2012)
8. C. J. Joachain
   Quantum Collision Theory.
   Reprint Edition: Elsevier Science, Amsterdam (1984)
9. H. Friedrich
   Scattering Theory.
   Springer Verlag, Heidelberg (2013)

On the theory of direct reactions in nuclear physics:

1. N. Oyster
   Direct Nuclear Reaction Theories.
   Wiley-Interscience, New York (1970).
2. G. R. Satchler
   Direct Nuclear Reactions.
   Clarendon Press, Oxford (1983)
3. N. K. Glendenning
   Direct Nuclear Reactions.
   Academic Press, New York (1983)
   New Edition: World Scientific, Singapore (2004)

Literature Cited in the Individual Chapters
Preliminary remarks and Chap. 1:

1. H. Geiger and E. Marsden, Phil. Mag. **25**, p. 604 (1913).
2. E. Rutherford, Phil. Mag. **21**, p.669 (1911).

1. N. Levinson, Danske Videnskab. Selskab, Mat.-fys. Medd. **25**, No 9 (1949).
2. S. Flügge, Practical Quantum Mechanics. SpringerVerlag, Heidelberg (1974)
3. M. Abramowitz, I. Stegun, Handbook of Mathematical Functions. Dover Publications, New York (1974)
4. P. Moon, D. Eberle, Field Theory Handbook. Springer Verlag, Heidelberg (1961)
5. J. R. Taylor: Scattering Theory, the Quantum Theory of Nonrelativistic Collisions. John Wiley, New York (1972)

Chapter 2

1. B. A. Lippmann, Phys. Rev. Lett. **79**, p. 461 (1950).
2. R. P. Feynman, Phys. Rev. **76**, p. 749 (1949).

Chapter 3

1. C. Møller, Danske Videnskab. Selskab, Mat-fys. Medd. **23**, p. 1 (1948).
2. J. A. Wheeler, Phys. Rev. **52**, p. 1107 (1937).
3. W. Heisenberg, Z. Phys. **120**, p. 513 (1943).

Chapter 4

1. M. E. Rose: Elementary Theory of Angular Momentum. J. Wiley, New York (1957). Reprinted by Dover Publications, New York (1995).
2. C. Itzykson and J. B. Zuber: Quantum Field Theory. McGraw-Hill, New York (1985).

Chapter 5

1. H. Poincaré, Acta Math. **4**, p. 201 (1884).
2. R. Jost, Helv. Physica Acta **20**, p. 256 (1947).
3. R. G. Newton, J. Math.Phys. **1**, p. 319 (1960).
4. S. T. Ma, Phys. Rev. **69**, p. 668 (1946) and **71**, p. 195 (1947).
5. H. M. Nussenzveig, Nucl. Phys. **11**, p. 499 (1959)
6. K. Knopp: Theory of Functions. Dover Publications, New York (1996)

Chapter 7

1. L. E. Espinola Lopez, J. J. Soares Neto, Int. J. Theor. Phys. **39**, p. 1129 (2000).
2. A. G. Sitenko: Lectures in Scattering Theory. Pergamon Press, Oxford (1971)
3. L. D. Faddeev, JETP **12**, p. 1014 (1961)
4. C. Lovelace, Phys. Rev. **135**, p. B1225 (1964).
5. W. Glöckle: The Quantum Mechanical Few-Body Problem. Springer Verlag, Heidelberg (1983)
6. E. O. Alt, P. Grassberger, W. Sandhas, Nucl. Phys. **B2**, p.167 (1967).
7. A. de Shalit and I. Talmi, Nuclear Shell Theory. Academic Press, New York (1963)
8. S. T. Butler, Phys. Rev. **80**, p. 1095 (1950), Proc. R. Soc. (London) **A208**, p. 559 (1951), and Phys. Rev. **88**, p. 685 (1952).
9. R. D. Woods and D. S. Saxon, Phys. Rev. **95**, p. 577 (1954).
10. P. E. Hodgson: The Nucleon Optical Model. World Scientific, Singapore (1994)

Recurring Literature

1. Abramowitz/Stegun:
   M. Abramowitz, I. Stegun: Handbook of Mathematical Functions. Dover Publications, New York (1974)
2. R. M. Dreizler, C. S. Lüdde: Theoretical Mechanics. Springer Verlag, Heidelberg (2002 and 2008)

Index

© Springer-Verlag GmbH Germany, part of Springer Nature 2022
R. M. Dreizler et al., *Quantum Collision Theory of Nonrelativistic Particles*,
https://doi.org/10.1007/978-3-662-65591-7

Printed in the United States
by Baker & Taylor Publisher Services